TIME-SERIES FORECASTING

TIME-SERIES
FORECASTING

Chris Chatfield

Reader in Statistics
Department of Mathematical Sciences
University of Bath, UK

CHAPMAN & HALL/CRC

Boca Raton   London   New York   Washington, D.C.

Library of Congress Cataloging-in-Publication Data

Catalog record is available from the Library of Congress.

© 2001 by Chapman & Hall/CRC

No claim to original U.S. Government works
International Standard Book Number 1-58488-063-5
Printed in the United States of America 1 2 3 4 5 6 7 8 9 0
Printed on acid-free paper

To

Steve and Sarah, Rob and Lali

wishing you fulfillment and contentment

Contents

Preface

Forecasting is an important activity in economics, commerce, marketing and various branches of science. This book is concerned with forecasting methods based on the use of time-series analysis. It is primarily intended as a reference source for practitioners and researchers in forecasting, who could, for example, be statisticians, econometricians, operational researchers, management scientists or decision scientists. The book could also be used as a text for a graduate-level course in forecasting. Some application areas, such as meteorology, involve specialist methods which are not considered in this book. The specialist area of judgemental forecasting is also not covered. Rather we concentrate on time-series forecasting methods which are applicable more generally, but especially in economics, government, industry and commerce.

The scope of the subject is wide and the topics I have chosen to cover reflect my particular interests and concerns. There are now several books, both at introductory and intermediate levels, which aim to cover the basic range of time-series forecasting methods. This book does not attempt to duplicate this material and is not meant to be a 'how-to-forecast' manual. It would need to be several times longer in order to provide all the necessary information about theory and applications for all the methods reviewed in this book. Rather this book was conceived as providing a summary of the basics of time-series modelling (Chapters 2 and 3), followed by a *fairly brief catalogue* of time-series forecasting methods, including recent arrivals such as GARCH models and neural networks (Chapters 4 and 5). Following this overview, the book attempts to *compare* the more important methods, both in terms of their theoretical relationships (if any, see Chapters 4 and 5), and with their practical merits including their empirical accuracy (Chapter 6). While the search for a 'best' method continues, it is now well established that no single method will outperform all other methods in all situations, and, in any case, it depends on what is meant by 'best'! The context is crucial.

This book also covers two other general forecasting topics, namely the *computation of prediction intervals* (Chapter 7) and the *effect of model uncertainty on forecast accuracy* (Chapter 8). These important aspects of forecasting have hitherto received rather little attention in the time-series literature. Point forecasts are still presented much more frequently than interval forecasts, even though the latter are arguably much more useful. There are various reasons for the overemphasis on point forecasts and it is hoped that the presentation in Chapter 7 will help to change the position. The model uncertainty problem arises when the analyst formulates *and* fits a model to the *same* set of data – as is the usual procedure in time-series analysis. The analyst typically behaves as if the selected model is the correct model and ignores the biases which arise when the same data are used to choose, fit and use a model. Standard least-squares theory no longer applies but is typically used anyway. Prediction intervals are typically calculated conditional on the fitted model and are often found to be too narrow. This provides a link with the material in Chapter 7.

While self-contained in principle for a reader with a reasonably thorough general statistical background, this book will be more accessible for someone who knows something about the basics of time-series analysis and has, for example, some understanding of trend, seasonality, autoregressive (AR) and moving average (MA) models.

I would like to acknowledge helpful comments on draft versions of the book from various people including Keith Ord, Sandy Balkin and two anonymous referees. I thank Ken Wallis for providing Figure 7.1. The work for Example 8.4 was carried out jointly with Julian Faraway (University of Michigan, USA). As always, any errors or obscurities that remain are my responsibility, however much I may wish to avoid this!

The time-series data used as examples in this book are available via my website http://www.bath.ac.uk/~mascc/ at the University of Bath. They include an extended S&P 500 series of 9329 observations from 3/07/1962 to 22/07/1999. I would like to have included more examples, but the manuscript is already long enough and time is pressing. My forecasted date of delivery has proved remarkedly inaccurate! I hope that you, the reader, find the book clear and helpful. Please feel free to tell me what you think as constructive criticism is always welcome.

Chris Chatfield
Department of Mathematical Sciences
University of Bath
Bath, U.K., BA2 7AY
email: cc@maths.bath.ac.uk
October 2000

Abbreviations and Notation

AR	Autoregressive
MA	Moving Average
ARMA	Autoregressive Moving Average
ARIMA	Autoregressive Integrated Moving Average
SARIMA	Seasonal ARIMA
ARFIMA	Fractionally Integrated ARIMA
TAR	Threshold Autoregressive
GARCH	Generalized Autoregressive Conditionally Hereoscedastic
SWN	Strict White Noise
UWN	Uncorrelated White Noise
NN	Neural Network
MSE	Mean Square Error
PMSE	Prediction Mean Square Error
MAPE	Mean Absolute Prediction Error
P.I.	Prediction Interval
ac.f.	Autocorrelation function
acv.f.	Autocovariance function
$\hat{x}_N(h)$	Forecast of x_{N+h} made at time N
B	Backward shift operator
∇	The operator $(1 - B)$
E	Expectation or Expected value
I	The identity matrix – a square matrix with ones on the diagonal and zeroes otherwise
$N(\mu, \sigma^2)$	Normal distribution, mean μ and variance σ^2
χ^2_ν	Chi-square distribution with ν degrees of freedom
Z_t or ε_t	Uncorrelated white noise with constant mean and variance

Vectors are indicated by boldface type, but not scalars or matrices.

CHAPTER 1

Introduction

"Don't never prophesy: If you prophesies right, ain't nobody going to remember and if you prophesies wrong, ain't nobody going to let you forget."

$-$ Mark Twain

Good forecasts are vital in many areas of scientific, industrial, commercial and economic activity. This book is concerned with *time-series forecasting*, where forecasts are made on the basis of data comprising one or more time series. A *time-series* is a collection of observations made sequentially through time. Examples include (i) sales of a particular product in successive months, (ii) the temperature at a particular location at noon on successive days, and (iii) electricity consumption in a particular area for successive one-hour periods. An example is plotted in Figure 1.1.

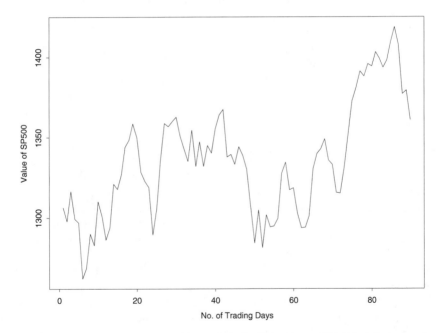

Figure 1.1 *A graph showing the Standard & Poor (S&P) 500 index for the U.S. stock market for 90 trading days starting on March 16 1999. (Note that values for successive trading days are plotted at equal intervals even when weekends or public holidays intervene.)*

Applications of time-series forecasting include:

1. Economic planning

2. Sales forecasting

3. Inventory (or stock) control

4. Production and capacity planning

5. The evaluation of alternative economic strategies

6. Budgeting

7. Financial risk management

8. Model evaluation

Most of the applications on the above list are self-explanatory. For example, good forecasts of future sales will obviously make it easier to plan production. However, the reader may not have realized that forecasts can help in *model evaluation*, when trying to fit time-series models. Checks on fitted models are usually made by examining the goodness-of-fit of the same data used to estimate the model parameters (called the *in-sample* fit). However, time-series data provide an excellent opportunity to look at what is called *out-of-sample* behaviour. A time-series model will provide forecasts of new future observations which can be checked against what is actually observed. If there is good agreement, it will be argued that this provides a more convincing verification of the model than in-sample fit; – see Sections 3.5.3 and 8.5.4.

The book does not attempt to cover the specialist areas of *population* and *weather forecasting*, although time-series techniques could, for example, be used to forecast a specific variable such as temperature. We also exclude specific mention of *probability forecasting*, where the aim is to predict the probability of occurrence of a specific event outcome such as a turning point, a change in interest rate or a strike.[1]

This book is not intended to be a comprehensive manual of forecasting practice, as several books are already available which describe in detail how to apply particular forecasting methods. Rather this book aims to give an overview of the many methods available, and to provide appropriate up-to-date references for the reader who wants more detail. There is emphasis on the intuitive ideas underlying the different methods, some comments on recent research results, and some guidance on coping with practical problems. The inter-relationships between the different methods are explored and the methods are compared from both theoretical and empirical points of view. After revising the basics of time-series analysis in Chapter 2, Chapter 3 discusses model building for a single series, both for particular classes of model and in general terms. Chapters 4 and 5 look at univariate and multivariate forecasting methods, respectively, while Chapter 6 discusses the evaluation of forecasts and gives advice on

---

[1] See, for example, the special issue on Probability Forecasting, *Int. J. of Forecasting*, 1995, No. 1.

which method to choose in different contexts. Chapters 7 and 8 cover two additional important topics, namely different ways of calculating prediction intervals (rather than point forecasts), and the effect of model uncertainty on forecast accuracy, especially in regard to the tendency for empirical results to suggest that prediction intervals are generally too narrow.

This opening introductory chapter begins by discussing the different categories of forecasting method in Section 1.1, while some preliminary practical questions are raised in Section 1.2. Section 1.3 contains a 'Public Health Warning' on the dangers of extrapolation, and Section 1.4 explains the important distinction between forecasts which are made 'in-sample' (and which are therefore not genuine forecasts) and those which are 'out-of-sample'. The chapter ends in Section 1.5 with a brief overview of relevant literature.

1.1 Types of forecasting method

Suppose we have an observed time series $x_1, x_2, \ldots, x_N$ and wish to forecast future values such as x_{N+h}. The integer h is called the *lead time* or the *forecasting horizon* (h for horizon) and the forecast of x_{N+h} made at time N for h steps ahead will be denoted by $\hat{x}_N(h)$. Note that it is essential to specify both the time the forecast is made *and* the lead time. Some of the literature does not do this and instead uses an ambiguous notation such as $\hat{x}_{N+h}$ for forecasts of X_{N+h} regardless of when the forecast was made.

A *forecasting method* is a procedure for computing forecasts from present and past values. As such it may simply be an algorithmic rule and need not depend on an underlying probability model. Alternatively it may arise from identifying a particular model for the given data and finding optimal forecasts conditional on that model. Thus the two terms 'method' and 'model' should be kept clearly distinct. It is unfortunate that the term 'forecasting model' is used rather loosely in the literature and is sometimes wrongly used to describe a forecasting *method*.

Forecasting methods may be broadly classified into three types:

(a) *Judgemental forecasts* based on subjective judgement, intuition, 'inside' commercial knowledge, and any other relevant information.

(b) *Univariate methods* where forecasts depend only on present and past values of the single series being forecasted, possibly augmented by a function of time such as a linear trend.

(c) *Multivariate methods* where forecasts of a given variable depend, at least partly, on values of one or more additional time series variables, called predictor or explanatory variables. Multivariate forecasts may depend on a multivariate model involving more than one equation if the variables are jointly dependent.

More generally a forecasting method could combine more than one of the above approaches, as, for example, when univariate or multivariate forecasts are adjusted subjectively to take account of external information which is difficult to express formally in a mathematical model.

This book focuses on univariate and multivariate time-series methods, and does not attempt to cover judgemental forecasting. Instead the reader is referred, for example, to the literature review by Webby and O'Connor (1996). The most famous judgmental method is probably that called the *Delphi technique* (e.g. Rowe and Wright, 1999), which aims to find a consensus of opinion for a group of 'experts', based on a series of questionnaires filled in by individuals and interspersed with the controlled feedback of opinion and information from other experts. Empirical results suggest that judgemental methods sometimes work well and sometimes do not (as is true for all forecasting methods!). However, statistical methods tend to be superior in general, provided that the latter are a practical proposition. Of course, there are occasions when model-based methods are not practical, perhaps because some essential information is not available, and then judgmental methods have to be used anyway. In any case, it is not sensible to pretend that the modelling and judgemental approaches are completely distinct, as it often helps to combine the two approaches and get the best of both worlds. In particular, many macroeconomic forecasts are obtained by making adjustments to model-based forecasts, perhaps by adding or subtracting an appropriate constant (sometimes called an intercept correction). It is, however, unfortunate that it is not always made clear how such adjustments are made in order to arrive at a final forecast. As might be expected, the integration of a judgemental and statistical approach can improve accuracy when the analyst has good domain knowledge but can harm accuracy when judgement is biased or unstructured (Armstrong and Collopy, 1998).

While the techniques commonly known as 'judgemental forecasting' are not covered in this book, it should be clearly understood that some element of judgement *is* always involved in forecasting, even when using what is normally regarded as an 'objective' statistical method. Good time-series modelling, like all statistical model building, involves the use of sound subjective judgement in assessing data, selecting a model, and interpreting the results. This use of judgement *will* be covered in what follows.

An alternative important way of classifying forecasting methods is between *automatic* methods, which require no human intervention, and *non-automatic* methods, which do. As one example of this important distinction, it is instructive to compare inventory control with economic planning. In inventory control there may be hundreds, or even thousands, of items to monitor. It is quite impossible to fit separate models to each individual time series of sales. Rather, a simple, automatic method is normally used for the whole range of items. In contrast, economic planning requires the analyst to carefully build an appropriate model describing the relationship between relevant economic variables, after which 'optimal' forecasts can be produced from the model. This method is certainly *not* automatic.

Yet another classification is between *simple* and *complicated* methods. Univariate methods are generally simpler than multivariate methods and it is always a matter of judgement as to how much effort should go into making a prediction. For some economic series, it is hard to beat the very simple forecast of using the latest observation to predict the next one. A deterministic series is also easy to forecast when the model is known. For example, a series which changes periodically through time in a systematic way is easy to forecast once a full cycle has been observed. At the other extreme, a series of independent observations is also 'easy' (or impossible) to predict as you cannot do better than use the overall mean value. Our main interest is in series which are in-between these extremes. They cannot be predicted exactly but they do contain structure which can be exploited to make better forecasts.

1.2 Some preliminary questions

In forecasting, as in any statistical exercise, it is essential to carry out any necessary preliminary work. In particular, it is important to *formulate the problem* carefully (see Chatfield, 1995a, Chapter 3). The analyst must (i) *ask questions* so as to get sufficient background information, (ii) *clarify the objectives* in producing forecasts, and (iii) find out exactly how the forecast will be used. The *context* is crucial in all of this. Ideally forecasts should be an integral part of the planning system and not a separate exercise. This is sometimes called a *systems approach*. It requires that the statistician talks to the people who will actually *use* the forecasts. A relatively simple forecasting method, which is widely understood, may be preferred.

A key related question is whether the forecasts actually influence the outcome. In some situations the forecast is used as a *target* value, while in others the forecasts are used to suggest control action. For example, a sales forecast may become a target in that workers will try to achieve sales of at least the forecast value, while a forecast of an increasing death rate for a particular disease may lead to preventive action to try to reduce the spread of the disease. Forecasts which prompt control action will be self-defeating, and yet such forecasts can be very useful even though they may not score well in terms of accuracy. In one sense, short-term weather forecasting is easier than econometric forecasting, as short-term weather forecasts cannot influence the weather, whereas economic forecasts may influence governmental policy. However, note that long-term weather forecasting (over a period of years, rather than days) could affect the outcome, in that predictions of global warming, for example, may influence government policy to try to reduce greenhouse gases, so as to prevent unwanted changes to weather patterns.

The analyst should also find out if forecasts only are required, or if there is a need for a descriptive, interpretable *model* and whether the forecasts are going to be used for control purposes. As always there is no point in giving the RIGHT answer to the WRONG question (called an *error of the third kind!*). For example, a model that gives a good *fit* to a set of past data may or may not be the most useful model for predicting future values. Fitting past values and forecasting future values are two quite different applications of a model. Similarly the best-fit model may or may not be the most helpful model for providing 'policy advice', such as deciding whether a proposed change in tax rates is likely to result in a beneficial effect on future unemployment. The sort of problem that arises in practice is that economic theory may suggest that one particular variable should be an important explanatory variable in an econometric model but, in a given set of recent data, it is found that the variable has been held more or less constant over the relevant period. Then time-series methods are likely to show that the inclusion or exclusion of the variable makes little difference to the overall fit of a model in regard to explaining the variation in the dependent (or response) variable. Furthermore, on the basis of the given data, it may be impossible to assess the effect of changes in the explanatory variable on forecasts of the response variable. Thus a model derived empirically will be useless for predicting changes in the response variable if a substantial change to the explanatory variable is envisaged, even though the analyst knows that the variable cannot be disregarded. Thus an appropriate model will need to be constructed using economic theory rather than (just) goodness-of-fit to past data.

The other main preliminary task is to make a careful assessment of the available data (sometimes called the *information set*). Have the 'right' variables been recorded to the 'right' accuracy? How much of the data is useful and usable? Are there obvious errors, outliers and missing observations? There is little point in putting a lot of effort into producing forecasts if the data are of poor quality in the first place.

The analyst may also have to ask various additional questions such as how many series are there to forecast, how far ahead are forecasts required, and what accuracy is desired in the forecasts (it may be wise to dampen unrealistic expectations). The forecaster must also decide whether to compute a point forecast, expressed as a single number, or an interval forecast. The latter is often more desirable, though rather neglected in the literature.

The answers to all the above questions determine, in part, which forecasting method should be chosen, as do more pragmatic considerations such as the skill and experience of the analyst and the computer software available. The analyst is advised to use a method he or she feels 'happy' with and if necessary to try more than one method. A detailed comparison of the many different forecasting methods is given in Chapter 6, based on theoretical criteria, on empirical evidence, and on practical considerations.

The importance of clarifying objectives cannot be overstressed. The forecasting literature concentrates on *techniques* – how to implement particular forecasting methods – whereas most forecasters probably need much more help with the *strategy* of forecasting. For example, there is a plenty of software available to make it easy to fit a class of time-series models called ARIMA models (see Section 3.1) but it is still hard to know *when* to use an ARIMA model and *how* to choose which ARIMA model to use.

1.3 The dangers of extrapolation

It is advisable to include here a brief warning on the dangers of forecasting. Time-series forecasting is essentially a form of extrapolation in that it involves fitting a model to a set of data and then using that model outside the range of data to which it has been fitted. Extrapolation is rightly regarded with disfavour in other statistical areas, such as regression analysis. However, when forecasting the future of a time series, extrapolation is unavoidable. Thus the reader should always keep in mind that forecasts generally depend on the future being like the past.

Forecasts also depend on the assumptions which are (implicitly or explicitly) built into the model that is used, or into the subjective judgements that are made. Thus forecasts are generally *conditional* statements of the form that "if such-and-such behaviour continues in the future, then ... ". It follows that one should always be prepared to modify forecasts in the light of any additional information, or to produce a range of different forecasts (rather than just one forecast) each of which is based on a known set of clearly stated assumptions. The latter is sometimes called *scenario forecasting* – see Section 8.5.2 – and Schumacher (1974) is correct when he says that "long-term feasibility studies, based on clearly stated assumptions, are well worth doing". However, Schumacher also says that long-term forecasts "are presumptious", but it is not clear to me when a "feasibility study" becomes a forecast. What we can say is that any really long-term forecast is liable to be way off target. Recent examples I have seen, which I expect to be wrong, include traffic flow in 2020 and world population in 2050! Fortunately for the forecaster, most people will have forgotten what the forecast was by the time it 'matures'. Celebrated examples, which have not been forgotten, include the founder of IBM predicting "a world market for about five computers" in 1947, and the President of Digital Equipment predicting that " there is no reason for any individual to have a computer in their home" in 1977. Going further back into history, I like the quote attributed to U.S. President Hayes in 1876, after he had witnessed a telephone call, that it was "an amazing invention but who would ever want to use one?"

Of course, forecasts can even go horribly wrong in the short-term when there is a sudden change or 'structural break' in the data – see Section 8.5.5. One famous example is that made by a Professor of Economics at Yale University in September 1929, when he said that "Stock prices have reached what looks like a permanently high plateau". This was just before the stock market 'crash', which led on to the Depression!

1.4 Are forecasts genuinely out-of-sample?

"Prediction is very difficult, especially if it's about the future" –   Nils Bohr

In a real forecasting situation, the analyst typically has data up to time N, and makes forecasts about the future by fitting a model to the data up to time N and using the model to make projections. If $\hat{x}_N(h)$ only uses information up to time N, the resulting forecasts are called *out-of-sample* forecasts. Economists call them *ex-ante* forecasts.

One difficulty for the analysts is that there is no immediate way of calibrating these forecasts except by waiting for future observations to become available. Thus it is sometimes helpful to check the forecasting ability of the model using data already at hand. This can be done in various ways. If the model is fitted to all the data and then used to 'forecast' the data already used in fitting the model, then the forecasts are sometimes called *in-sample forecasts*, even though they are not genuine forecasts. The one-step-ahead in-sample forecasts are in fact the *residuals* – see Section 3.5.3. An alternative procedure (see Section 8.5.4) is to split the data into two portions; the first, sometimes called the *training set*, is used to fit the model, while the second portion, sometimes called the *test set*, is used for calibration purposes to check forecasts made by the model. The properties of the resulting forecasts are more like those of real forecasts than the residuals. Of course, if one really believes one has the 'true' model, then the properties of residuals and forecast errors should be similar, but, we see in later chapters, that, in practice, out-of-sample forecasts are generally not as accurate as would be expected from in-sample fit. Fortunately, many of the comparative forecasting studies, which have been reported in the literature, do use a test set for making comparisons. Even so, there are many ways in which forecasts can be 'improved' by procedures which are of dubious validity and the reader is strongly advised to check that results on comparative forecast accuracy really do relate to forecasts made under similar conditions. In particular, if all forecasts are meant to be out-of-sample, then the different forecasting methods being compared should only use historical data in the information set.

There are several ways in which forecasts can be unfairly 'improved'. They include:

1. Fitting the model to all the data including the test set.

2. Fitting several different models to the training set, and then choosing the model which gives the best 'forecasts' of the test set. The selected model is then used (again) to produce forecasts of the test set, even though the latter has already been used in the modelling process.

3. Using the known test-set values of 'future' observations on the explanatory variables in multivariate forecasting. This will obviously improve forecasts of the dependent variable in the test set, but these future values will not of course be known at the time the forecast is supposedly made (though in practice the 'forecast' is made at a later

date). Economists call such forecasts *ex-post* forecasts to distinguish them from *ex-ante* forecasts. The latter, being genuinely out-of-sample, use forecasts of future values of explanatory variables, where necessary, to compute forecasts of the response variable – see also Section 5.1.2. *Ex-post* forecasts can be useful for assessing the effects of explanatory variables, provided the analyst does not pretend that they are genuine out-of-sample forecasts.

In my experience, it often seems to be the case that, when one method appears to give much better forecasts than alternatives, then the 'good' method has some unfair advantage. It is therefore unfortunate that some published empirical studies do not provide sufficient information to see exactly how forecasts were computed, and, in particular, to assess if they are genuinely out-of-sample. When this happens, the suspicion remains that the results are compromised. Of course, there will sometimes be good reasons for computing alternatives to genuine out-of-sample forecasts. For example, in scenario forecasting, the analyst wants to assess the effect of making different assumptions about future values of explanatory variables. This sort of exercise is perfectly reasonable provided that the assumptions are clearly stated and the forecaster realizes that it will not be fair to compare such results with forecasts based solely on past data.

1.5 Brief overview of relevant literature

This section gives a brief review of books on time-series analysis and forecasting. Some help with scanning research-level journals for articles (papers) on forecasting is also given. Additional references to more specialized books and papers are given throughout the book.

General introductory books on time-series analysis include Brockwell and Davis (1996), Chatfield (1996a), Diggle (1990), Harvey (1993), Janacek and Swift (1993), Kendall and Ord (1990) and Wei (1990). More advanced books include the comprehensive two-volume treatise by Priestley (1981) which is particularly strong on spectral analysis, multivariate time series and non-linear models. The fourth edition of Volume 3 of Kendall and Stuart (Kendall, Stuart and Ord, 1983) is also a valuable reference source, but earlier editions are now rather dated. Other intermediate to advanced books include Anderson (1971), Brillinger (1981), Brockwell and Davis (1991) and Fuller (1996). The books by Enders (1995), Hamilton (1994), Harvey (1990) and Mills (1990, 1999) are more suitable for the reader with an econometric background.

The famous book by Box and Jenkins (1970) describes an approach to time-series analysis, forecasting and control which is based on a class of linear stochastic processes, called ARIMA models. The revised edition published in 1976 was virtually unchanged, but the third edition (Box et al., 1994) with G. Reinsel as co-author is a substantial revision of earlier editions. In particular, Chapters 12 and 13 have been completely rewritten and include new material on topics such as intervention analysis, outlier detection and process control. We therefore generally refer to the new third edition. However, readers with an earlier edition will find that Chapters 1 to 11 of the new edition retain the spirit and a similar structure to the old one, albeit with some revisions such as new material on ARMA model estimation and on testing for unit roots. This book covers ARIMA models in Section 3.1 and gives a brief description of the Box-Jenkins approach in Section 4.2.2. Readers with little experience of ARIMA modelling, who want further details, may be advised to read Vandaele (1983), or one of the introductory time-series texts given above, rather than Box et al. (1994).

There are a number of books which are targeted more towards forecasting, rather than general time-series analysis. Granger and Newbold (1986) is a good general book on the topic, especially for applications in economics. Some other general texts on time-series forecasting include Abraham and Ledolter (1983) and Montgomery et al. (1990). The texts by Bowerman and O'Connell (1987), Diebold (1998), Franses (1998), and Makridakis et al. (1998) are aimed more at business and economics students. There is a useful collection of up-to-date review articles in Armstrong (2001). Some important, more specialized, books include Harvey's (1989) book on *structural models* and West and Harrison's (1997) book on *dynamic linear models*, which is written from a Bayesian viewpoint. Pole et al. (1994) give some case studies using the latter approach together with a software package called BATS.

The two main journals devoted to forecasting are the *International Journal of Forecasting* (sponsored by the International Institute of Forecasters and published by North-Holland) and the *Journal of Forecasting* (published by Wiley). Papers on forecasting can also appear in many other statistical, management science, econometric and operational research journals including the *Journal of Business and Economic Statistics*, *Management Science* and the *Journal of Econometrics*. A brief general review of recent developments in time-series forecasting is given by Chatfield (1997). Keeping up with the literature in an interdisciplinary subject like forecasting is difficult. Abstract journals may help and 'word of mouth' at specialist conferences is also invaluable.

CHAPTER 2

Basics of Time-Series Analysis

This chapter gives a brief review of some basic tools and concepts in time-series analysis. This will be useful for reference purposes, to set up our notation, and to act as revision material for readers who are hopefully familiar with some of these ideas.

2.1 Different types of time series

A *time series* is a set of observations measured sequentially through time. These measurements may be made continuously through time or be taken at a discrete set of time points. By convention, these two types of series are called *continuous* and *discrete* time series, respectively, even though the measured variable may be discrete or continuous in either case. In other words, for discrete time series, for example, it is the time axis that is discrete.

For a continuous time series, the observed variable is typically a continuous variable recorded continuously on a trace, such as a measure of brain activity recorded from an EEG machine. The usual method of analysing such a series is to sample (or digitize) the series at equal intervals of time to give a discrete time series. Little or no information is lost by this process provided that the sampling interval is small enough, and so there is no need to say anything more about continuous series.

In fact, discrete time series may arise in three distinct ways, namely

1. by being *sampled* from a continuous series (e.g. temperature measured at hourly intervals. Such data may arise either by sampling a continuous trace, as noted above, or because measurements are only taken once an hour);

2. by being *aggregated* over a period of time (e.g. total sales in successive months);

3. as an inherently discrete series (e.g. the dividend paid by a company in successive years).

For all three types of discrete time series, the data are typically recorded at *equal intervals of time*. Thus the analysis of equally spaced discrete time series constitutes the vast majority of time-series applications and this book restricts attention to such data. The treatment of the three types of discrete series is usually very similar, though one may occasionally wish to consider the effect of using different sampling intervals for continuous series or different periods of aggregation for aggregated data.

It is worth noting that data may be *aggregated* either across time *or* across series, and these two situations are quite different. The first is called *temporal aggregation* and the second is called *contemporaneous aggregation*. For example, suppose we have sales figures for each of the many different brand sizes of different products in successive weeks. Such data may be quite volatile and difficult to forecast without some form of aggregation, either across time (e.g. over successive 4-week periods) or across products (e.g. sum all brand sizes for the same brand).

The possibility of aggregating data raises many questions such as how to choose the best level of aggregation for making forecasts, and how to use monthly data to improve quarterly forecasts (e.g. Montgomery et al., 1998). For example, a common problem in inventory control is whether to develop a summary forecast for the aggregate of a particular group of items and then allocate this forecast to individual items based on their historical relative frequency – called the *top-down* approach – or make individual forecasts for each item – called a *bottom-up* approach. Dangerfield and Morris (1992) present empirical results which suggest the bottom-up approach tends to be more accurate, but these are difficult and rather specialist problems which we do not address directly in this book. I have seen rather little general advice in the literature on aggregation, partly because (as is often the case in forecasting!) there are no general guidelines, and so contextual considerations are often paramount.

2.2 Objectives of time-series analysis

Suppose we have data on one or more time series. How do we go about analysing them? The special feature of time-series data is that successive observations are usually *not* independent and so the analysis must take account of the *order* in which the observations are collected. Effectively each observation on the measured variable is a bivariate observation with time as the second variable.

The main *objectives* of time-series analysis are:

(a) *Description.* To describe the data using summary statistics and/or graphical methods. A time plot of the data is particularly valuable.

(b) *Modelling.* To find a suitable statistical model to describe the data-generating process. A *univariate* model for a given variable is based only on past values of that variable, while a *multivariate* model for a given variable may be based, not only on past values of that variable, but also on present and past values of other (predictor) variables. In the latter case, the variation in one series may help to *explain* the variation in another series. Of course, all models are approximations and model building is an art as much as a science – see Section 3.5.

(c) *Forecasting.* To estimate the future values of the series. Most authors use the terms 'forecasting' and 'prediction' interchangeably and we follow this convention. There is a clear distinction between steady-state forecasting, where we expect the future to be much like the past, and *What-if* forecasting where a multivariate model is used to explore the effect of changing policy variables.

(d) *Control.*   Good forecasts enable the analyst to take action so as to control a given process, whether it is an industrial process, or an economy or whatever. This is linked to What-if forecasting.

This book is primarily concerned with objective (c), but description and modelling are often a prerequisite. Thus the objectives are interlinked.

2.3 Simple descriptive techniques

Before trying to model and forecast a given time series, it is desirable to have a preliminary look at the data to get a 'feel' for them and to get some idea of their main properties. This will be invaluable later on in the modelling process. The time plot is the most important tool, but other graphs and summary statistics may also help. In the process, the analyst will also be able to 'clean' the data (see Section 2.3.4) by removing or adjusting any obvious errors. This whole approach is sometimes described as *Initial Data Analysis* (or IDA).

This section also describes some other simple descriptive methods which are specifically designed for analysing time-series data. The two main sources of variation in many time series are *trend* and *seasonal variation*, and they are considered in Section 2.3.5 and 2.3.6, respectively.

Little prior knowledge of time-series models is required throughout the section, but the ideas which are introduced should enhance the reader's general understanding of time-series behaviour before we go on to describe various classes of models.

It may help at this stage to present a brief summary of classical time-series analysis which aims to decomposes the variation in a time series into components due to:

(a) *Seasonal variation.*   This type of variation is generally annual in period and arises for many series, whether measured weekly, monthly or quarterly, when similar patterns of behaviour are observed at particular times of the year. An example is the sales pattern for ice cream which is always high in the summer. Note that if a time series is only measured annually (i.e. once per year), then it is not possible to tell if seasonal variation is present.

(b) *Trend.*   This type of variation is present when a series exhibits steady upward growth or a downward decline, at least over several successive time periods. An example is the behaviour of the U.K. retail price index which has shown an increase every year for many years.[1] Trend may be loosely defined as "long-term change in the mean level", but there

---

[1] Except when re-indexed to 100. I have seen at least one analysis which, sadly, did not take account of the discontinuity which arises when re-indexing takes place.

is no fully satisfactory mathematical definition. We will see that the perception of trend will depend, in part, on the length of the observed series. Furthermore, the decomposition of variation into seasonal and trend components is not unique.

(c) *Other cyclic variation.* This includes regular cyclic variation at periods other than one year. Examples include business cycles over a period of perhaps five years and the daily rhythm (called diurnal variation) in the biological behaviour of living creatures.

(d) *Irregular fluctuations.* The phrase 'irregular fluctuations' is often used to describe any variation that is 'left over' after trend, seasonality and other systematic effects have been removed. As such, they may be completely random in which case they cannot be forecast. However, they may exhibit short-term correlation (see below) or include one-off discontinuities.

Classical methods work quite well when the variation is dominated by a regular linear trend and/or regular seasonality. However, they do not work very well when the trend and/or seasonal effects are changing through time or when successive values of the irregular fluctuations are correlated. Correlation between successive values of the same time series is generally called *autocorrelation*. It is often found that successive residuals from a trend-and-seasonal model are correlated when separated by a short time interval and this is called short-term (auto)correlation. Then a more sophisticated modelling approach may well be desirable to improve forecasts.

2.3.1 *The time plot*

The first step in any time-series analysis or forecasting exercise is to plot the observations against time, to give what is called a *time plot* of the data. The graph should show up important features of the data such as trend, seasonality, outliers, smooth changes in structure, turning points and/or sudden discontinuities, and is vital, both in describing the data, in helping to formulate a sensible model and in choosing an appropriate forecasting method.

The general principles for producing a clear graph are covered in many books (e.g. Chatfield, 1995a) and include giving a graph a clear, self-explanatory title, choosing the scales carefully and labelling axes clearly. Even so, drawing a time plot is not as easy as it may sound. For example, it is not always obvious how to choose suitable scales or how to actually plot the points (e.g. using dots joined by straight lines?). Some software packages produce surprisingly poor graphs and yet may not allow the user the flexibility of making alternative choices of the presentational details, such as how the axes should be labelled. Even with a good package, some trial-and-error may still be necessary to get a graph suitable for the task at hand.

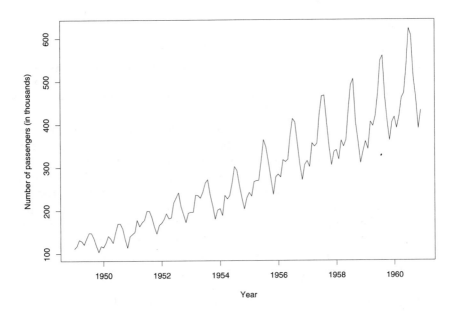

Figure 2.1. *The time plot of the Box-Jenkins airline data: Monthly totals, in thousands, of the numbers of international airline passengers from January 1949 to December 1960.*

Figure 2.1 shows a time plot of the airline passenger data listed by Box et al. (1994, Series G) produced by a software package called S-PLUS which does allow the user to choose sensible scales and label them properly, to put in a clear title, to join up the points or not, and so on. However, the reader should note that the version of the graph given here was not produced first time, but rather required several iterations to improve it to an acceptable standard of presentation. Generally speaking, more effort is needed to produce a 'good' graph than some analysts expect. Figure 2.1 shows a clear upward trend from year to year (numbers of passengers are generally increasing) together with seasonal variation within a year (more passengers travel in the summer).

Having emphasized the importance of the time plot, it should also be said that there are occasions when such a graph may be difficult to interpret. This may simply arise from a poor choice of scales and so it is worth stressing that a bad graph can be worse than no graph. However, at a more sophisticated level,[2] even a well-produced time plot may not make

---

[2] These comments presuppose some basic knowledge of time-series models

it easy discern whether a series is generated by a model which is (i) non-stationary (see Section 2.4 for a definition of stationarity) or (ii) 'nearly' non-stationary (see Section 3.1.9), nor whether a series is (iii) linear or (iv) non-linear (see Section 3.4).

2.3.2 Other graphics

The use of graphics in time-series analysis is a rather neglected topic in time-series texts – Diebold (1998, Chapter 3) is an exception. This is a pity because the human eye is very efficient at finding patterns and spotting anomalies in graphs.

The time plot is by far the most useful graphic, but the time ordering of data can make it difficult to adapt other types of graph for use with time series. For example, the *histogram*, which is a common type of graph in general statistics for showing the distribution of data, is of limited value for interpreting most time-series data, because trend and cyclic variation will typically distort any distributional behaviour. It can, however, be useful to plot the histogram of values from a detrended, deseasonalized series.

Another well-known type of graph is the *scatter plot*, which is used to explore the relationship between two variables. The time plot of a single variable can be regarded as a form of scatter plot with time being treated as the second variable. However, the scatter plot does not naturally cope with exploring the relationship between two time-ordered variables, where there are effectively four variables, namely the two measured variables and their corresponding sets of time points. Indeed, plotting multivariate time-series data, so as to reveal inter-relationships between the variables, is a difficult operation which preferably requires some practical experience. We defer further consideration of this topic until Section 5.1.

Another important type of diagnostic graph is the so-called *correlogram*. This will be discussed in Section 2.6 after Section 2.4 has introduced a function called the *autocorrelation function* for a very general class of models called stationary stochastic processes.

2.3.3 Transformations

A time plot may also help to decide if any variables need to be transformed prior to the main analysis. For example, if there is an upward trend in the time plot, and the variance appears to be increasing with the mean level, then a transformation may be thought desirable in order to stabilize the variance (or make it homoscedastic). A transformation may also be indicated to make the data more normally distributed if the observations appear to be skewed (e.g. with more 'spikes' upwards than down). Finally, if the seasonal effect appears to be multiplicative (see section on seasonal variation below), then it may be thought desirable to transform the data to make the seasonal effect additive, as linear effects are generally easier to handle.

One general class of transformations is the *Box-Cox transformation*. Given an observed time series $\{x_t\}$, the transformed series is given by

$$y_t = \begin{cases} (x_t^\lambda - 1)/\lambda & \lambda \neq 0 \\ \log x_t & \lambda = 0 \end{cases} \qquad (2.3.1)$$

where λ denotes the transformation parameter which may be estimated by subjective judgement or by a formal statistical procedure. When $\lambda \neq 0$, the Box-Cox transformation is essentially a power transformation of the form x_t^λ as the rest of the expression is introduced to make y_t a continuous function of λ at the value $\lambda = 0$.

Transformations are widely used, but there is, however, little evidence that the use of a non-linear transformation improves forecasts (Nelson and Granger, 1979) and there can be difficulties in interpreting a model for a transformed series. Moreover, when forecasts are transformed back to the original scale of measurement, they will generally be biased if the transformed forecasts are unbiased. My personal preference is to avoid transformations whenever possible and work with the observed data, when the fitted model and forecasts are readily interpretable. Exceptions arise when a transformed variable has a direct physical interpretation, as, for example, when percentage increases are of interest indicating a logarithmic transformation. A transformation is also advisable when asymmetry and/or changes in variance are severe.

2.3.4 Cleaning the data

An important part of the initial examination of the data is to assess the quality of the data and consider modifying them, for example, by removing any obvious errors. This process is often called *cleaning* the data, or *data editing*, and is an essential precursor to attempting to model data. Data cleaning could include modifying outliers, identifying and correcting obvious errors, and filling in (or imputing) any missing observations. The analyst should also deal with any other known peculiarities such as a change in the way that a variable is defined during the course of the data-collection process. Some action may be necessary before doing anything else, such as trying to fill in missing observations. However, some cleaning may follow a simple preliminary descriptive analysis, and, in time-series analysis, the construction of a time plot for each variable should reveal any oddities such as outliers and discontinuities.

An adequate treatment of time-series outliers is particularly important, and some useful references include Box et al. (1994, Section 12.2), Chen and Liu (1993) and Ledolter (1989). It turns out that *additive outliers*, that only affect a single observation, are more harmful than so-called *innovation outliers*, that affect all subsequent observations after the first time they appear through the dynamics of the system as expressed in the model. It can be difficult to tell the difference between outliers caused by errors, by non-linearity (see Section 3.4) and by having a non-normal 'error' distribution. *Robust* methods downweight the effect of extreme observations and are increasingly used. Interval forecasts can also be modified to cope with outliers – see Section 7.7.

The *context* is crucial in deciding how to modify data, if at all. This explains why it is essential to get background knowledge about the problem, and in particular to clarify the objectives. A corollary is that it is difficult to make any general remarks or give any general recommendations on data cleaning. We will simply say that it is essential to combine statistical theory with sound common sense and knowledge of the particular problem being tackled.

2.3.5 Trend

The novice time-series analyst may think that the phenomenon known as *trend* is relatively easy to understand. However, further thought will suggest that it is not, in fact, easy to define 'trend'. One possibility is to describe it as the "long-term change in the underlying mean level per unit time". This may appear reasonable, but what is meant by 'long-term' and what is 'the current mean level'? Our perception of trend, and our understanding of 'long-term', depends partly on the length of the observed time series. Thus current changes in climate, such as *global warming*, may be described as a trend over the period of a single person's lifetime, even though, over a very long period of perhaps several hundred years, they may turn out be cyclic variation at a low-frequency (though I wouldn't count on it!).

We now present various mathematical models to describe different forms of trend, both linear and non-linear, and both stochastic and deterministic. Let μ_t denote the local mean level at time t. Many older textbooks consider the case where

$$\mu_t = \alpha + \beta t \qquad\qquad (2.3.2)$$

This simple linear trend is often called a *deterministic* or *global* linear trend. However, current thinking (e.g. Newbold, 1988; Chatfield, 1996a, Chapter 10) is to generally avoid models including a deterministic function of time and to favour *local* as opposed to global, models. Thus the trend is modelled as evolving through time in a stochastic, rather than determnistic, way. One way of representing a local linear trend is to allow the parameters α and β in 2.3.2 to evolve through time so that

$$\mu_t = \alpha_t + \beta_t t \qquad\qquad (2.3.3)$$

where α_t denotes the local intercept, and β_t denotes the local slope. This type of model is used much more commonly nowadays, because it has been found that a deterministic linear trend rarely provides a satisfactory model for real data. For example, Franses and Kleibergen (1996) show that out-of-sample forecasts based on modelling the first differences of economic data are generally better than those obtained by fitting a deterministic trend. Computer software now exists to fit local trends and the latter seem to be generally more robust.

Before we go any further, it is important to realize that there is a potential for confusion in the literature in regard to the use of the term 'trend'. Some authors refer to μ_t in (2.3.2) and (2.3.3) as being the trend term, while others use the term 'trend' to describe the *rate of change* in μ_t, namely the slope, β or β_t. These quantities are quite different.[3] This book generally refers to μ_t as being the local *level*, and sometimes uses the term 'trend' to refer to the *rate of change* in μ_t, namely β or β_t. However we also use the term 'trend' as in normal parlance to mean any general tendency for the level to change through time and so generally use the phrase *growth rate* to describe the slope, β or β_t, in the hope that this will avoid any confusion.

An alternative way of modelling a local trend is to use a recursive equation as in state-space or structural models which are described more fully in Section 3.2. The global trend modelled by (2.3.2) may be rewritten as

$$\mu_t = \mu_{t-1} + \beta$$

and this recursive form of the equation is more convenient for many purposes. If we allow the change in level, namely β, to change through time and add a disturbance term, say $\{w_{1,t}\}$, then an alternative way of representing a *local* linear trend is by

$$\mu_t = \mu_{t-1} + \beta_{t-1} + w_{1,t} \tag{2.3.4}$$

to which must be added a second equation describing the evolution of $\{\beta_t\}$. The latter can be modelled in various ways. The most common assumption, described more fully in Section 3.2 in regard to what is called the *linear growth model*, is that $\beta_t = \beta_{t-1} + w_{2,t}$, where $\{w_{2,t}\}$ denotes a second disturbance process. We will see in Section 2.5.2 that this model for updating β_t is usually called a *random walk*. A recursive form of equation, such as (2.3.4), is generally used in preference to an equation like (2.3.3). Note that the level and growth rate, namely μ_t and β_t, are not observable directly and we will see that the linear growth model is completed by assuming that the observed value of the given time series is equal to $(\mu_t + n_t)$, where n_t denotes measurement error.

In many cases the trend is clearly *non-linear*. For example, the local level, μ_t, may be a polynomial function of time (such as a quadratic), or may show exponential growth (which can be difficult to handle, even if logarithms are taken so as to transform the trend to linearity), or may follow an S-shaped function of time such as the *Gompertz* curve

$$\log \mu_t = a + br^t \tag{2.3.5}$$

with $0 < r < 1$, or the *logistic* curve

$$\mu_t = a/(1 + be^{-ct}) \tag{2.3.6}$$

---

[3] Even more confusingly, many governmental statistics agencies refer to μ_t as being the *trend-cycle*. I think this term is very misleading.

The trend may be estimated in two quite distinct ways. Firstly, it is possible to fit the data with one of the parametric functions mentioned above, namely a polynomial function, a Gompertz curve or a logistic curve. Secondly, it is possible to apply a linear filter to the data in the form

$$\tilde{x}_t = \Sigma a_r x_{t+r} \qquad (2.3.7)$$

If $\Sigma a_r = 1$, then $\tilde{x}_t$ may be regarded as a smoothed value of x_t, say $\mathrm{Sm}(x_t)$. The difference between x_t and its smoothed value will then estimate the deviation from the trend and is a way of producing a detrended series. Note that $[x_t - \mathrm{Sm}(x_t)]$ is also a linear filter, of the form (2.3.7), but that the sum of the coefficients, Σa_r, is now zero, rather than one. The simplest example of a detrending linear filter, with Σa_r equal to zero, is that obtained by taking first differences of a series, namely $\nabla x_t = (x_t - x_{t-1})$. The reader should check that this is indeed of the form (2.3.7) with $a_0 = 1$ and $a_{-1} = -1$.

Many more complicated filters are available for measuring and/or removing trend. One important example is the *Henderson* family of moving averages which are designed to follow a cubic polynomial trend through time without distortion (Kenny and Durbin, 1982). It is possible to find a symmetric filter with this property covering different numbers of observations, with 9-, 13- or 23-term filters being particularly useful. A longer filter should be used when there are judged to be higher than average irregular fluctuations, or the data are monthly rather than quarterly. As one example, it can be shown that the symmetric 9-term Henderson filter has weights $(-0.041; -0.010; 0.119; 0.267; 0.330)$ with the last term being the centre term. The reader is invited to check that the sum of all nine coefficients is indeed one so that this filter is for trend estimation rather than trend removal. Of course, by subtracting each smoothed value from the corresponding observation, a detrended value will result.

The discussion of Ball and Wood (1996) demonstrates how difficult it still is to distinguish between different types of trend but we defer further discussion of this topic, including testing for unit roots, until Chapter 3.

It is also important to realize that the treatment of trend depends on whether or not seasonality is present. If seasonality *is* present, the analyst has to decide whether to measure and/or remove the seasonality before or after measuring the trend. In practice it is usual to adopt an iterative procedure. Preliminary estimates of trend and seasonality are found, typically with a fairly simple moving average, and these estimates are then revised using a more sophisticated smoother, until more refined estimates are obtained. Thus the treatment of trend and seasonality are inextricably related, reflecting the fact that there is no unique decomposition of variation into trend and seasonality. This leads on naturally to a more in-depth study of the topic of seasonal variation.

2.3.6 Seasonal variation

We consider variation over a period of one year, though results can readily
be adapted to other periods (e.g. variation on different days of the week).
The seasonal effect at time t will be denoted by i_t and is called the *seasonal
index* at time t. The seasonality is said to be *additive* when it does not
depend on the local mean level, and is said to be *multiplicative* when the
size of the seasonal variation is proportional to the local mean level. For
example, the seasonal variation in Figure 2.1 appears to be multiplicative,
rather than additive, as it becomes larger towards the end of the series.
In the additive case, the seasonal indices are usually normalized so that
$\Sigma i_t = 0$ (where the sum is over one year's values), while in the multiplicative
case, normalization can be achieved by adjusting the average value of (i_t)
over one year to equal one. If the seasonal and 'error' terms are both
multiplicative so that the observed random variable at time t, say X_t,
may be modelled by

$$X_t = \mu_t \, i_t \, \varepsilon_t \qquad (2.3.8)$$

where ε_t denotes the random disturbance at time t, then a logarithmic
transformation will turn this into the fully additive model

$$\log X_t = \log \mu_t + \log i_t + \log \varepsilon_t \qquad (2.3.9)$$

which may be easier to handle. Of course, (2.3.8) only makes sense if ε_t is
a non-negative random variable which has an expected value close to, or
equal to, one, and it is sometimes more instructive to rewrite (2.3.8) in the
form

$$X_t = \mu_t \, i_t \, (1 + \varepsilon_t') \qquad (2.3.10)$$

where $\varepsilon_t' = (\varepsilon_t - 1)$ has a mean close to, or equal to, zero.

There are a variety of methods for measuring and/or removing
seasonality. If you can carry out one of these operations (e.g. remove
seasonality), then you can generally do the other (e.g. estimate seasonality)
very easily as a result. For example, the difference between an observation
and its seasonally adjusted value will provide an estimate of the seasonal
effect in that period.

If the seasonal effect is thought to be constant through time, one
approach is to fit seasonal dummy variables using a regression approach
(e.g. Diebold, 1998, Chapter 5). This effectively fits one parameter for each
period during a year. Alternatively, if variation is thought to be smooth
through the year (e.g. sinusoidal), then it may be possible to model the
seasonality adequately using fewer parameters.

An alternative approach is to use some sort of linear filter, as in (2.3.7),
and various possibilities are available. *Seasonal differencing* is one very
simple way of deseasonalizing data. For example, with monthly data, where
there are 12 observations per year, the seasonal difference is written as
$\nabla_{12} x_t = (x_t - x_{t-12})$. Note that the sum of the coefficients, Σa_r, is zero, so
that this filter is removing seasonality, rather than estimating it. Seasonal

differencing is useful for building seasonal ARIMA models (see Section 3.1.6) but more sophisticated filters are needed for other purposes.

The measurement and removal of seasonal variation is complicated by the possible presence of *calendar* (or *trading day*) effects. The latter are the effects due to changes in the calendar, as for example when Easter falls in March one year and in April the next. Another example arises when a month contains five Sundays one year, but only four the next, so that the number of days when trading can occur will differ. This sort of event can make a large difference to some measures of economic and marketing activity. A review of the problem, which can be very important in practice, is given for example by Bell and Hillmer (1983). Many packages now allow such effects to be taken care of.

The U.S. Census Bureau devised a general approach to seasonal adjustment, called the *X-11 method*, which has been widely used in government statistical organizations to identify and remove seasonal variation. A recursive approach is used. Preliminary estimates of trend are used to get preliminary estimates of seasonal variation, which in turn are used to get better estimates of trend and so on. In addition, possible outliers are identified and adjusted during this process. The seasonality may be specified as additive or multiplicative. The user is able to select one of a number of linear filters, including a simple centred moving average (usually used at the first stage to get a preliminary estimate of trend) and Henderson filters[4] of different lengths. First published in 1967, the method has developed over the years, giving more options to the user in regard to the treatment of features such as outliers, calendar effects and sudden changes in the trend or seasonality.

When using a seasonal adjustment procedure as a precursor to forecasting, an important aspect of linear filters is what happens at the end of the series. Many linear filters, such as the Henderson trend-estimation filters, are symmetric, in that the smoothed value in a given period is obtained as a weighted average of past *and* future observations. Unfortunately, future values will not be available near the end of the series. This leads to *end effects* in that the filter cannot be used directly on the data. Asymmetric one-sided filters can be found which only use past observations, but they will typically make assumptions about the behaviour of future observations, as for example that they will project in a linear way in future.

A major development of the X-11 method was made by Statistics Canada during the 1980s with a seasonal adjustment program called X-11-ARIMA. This allows the user to augment the observed series by forecasting future values with a model called an ARIMA model, which will be described in Section 3.1.5. It also allows the user to backcast[5] extra values before the start of the series. Seasonal adjustment can then be readily carried out

---

[4] See Section 2.3.5.
[5] Backcasting means to forecast backwards.

by applying symmetric linear filters to the augmented series. The package allows the user to put in 'prior adjustments' to cope with features such as outliers, trend breaks (also called level shifts) and changes in the seasonal pattern. These prior adjustments are called 'temporary' if they are reversed at the end of the seasonal adjustment process so as to reflect a real effect in the data. They are called 'permanent' if they are not reversed, as for example with an outlier that is thought to be an error. Various options are available, such as ensuring that annual totals are constrained so that seasonally adjusted data have the same annual total as the original data. Aggregated series can be similarly handled.

The U.S. Census Bureau has recently developed an updated version of the X-11-ARIMA seasonal adjustment program, called X-12-ARIMA (see Findley et al., 1998). Its enhancements include more flexible capabilities for handling calendar effects, better diagnostics and an improved user interface, especially for handling large numbers of series. Like X-11-ARIMA, X-12-ARIMA has a very wide range of options, which should be seen as a strength, but could also be a weakness in the wrong hands. Clearly, users need adequate training. There are defaults, but they may not necessarily be sensible ones, and so it is not advised to use these packages in 'black-box' mode. Rather, the user is advised to get as much background knowledge as possible and have a preliminary look at the data (as in all time-series analysis).

X-11-ARIMA and X-12-ARIMA are widely used in economics and government, especially in Canada, the USA and the UK. In mainland Europe, some government statistics organizations use an alternative approach, based on a package called SEATS (Signal Extraction in ARIMA time series), that is described in Gomez and Maravall (2000). This model-based approach relies on seasonal ARIMA models (see Section 3.1.6), and involves pre-adjusting the data to cope with missing observations, outliers and calendar effects using another program called TRAMO (Time-series regression with ARIMA noise).

Many government statistics are presented in deseasonalized form so as to better assess the underlying trend, which is generally of prime interest. However, as noted earlier, there is no unique decomposition of variation into trend and seasonality, and it is sometimes found that data, thought to be seasonally adjusted, still contain some residual seasonality. If, instead, more operations are carried out on data so as to completely remove *all* seasonality, then it is sadly possible to distort other properties of the data such as the trend – see Gooijer and Franses (1997) for references – or even introduce non-linearities (Ghysels et al., 1996). Of course, the seasonal patterns are sometimes of interest in their own right and then it could be a mistake to deseasonalize the data except insofar as this leads to better estimates of the seasonality.

More information on seasonal adjustment is given by Butter and Fase (1991) and in the collection of papers from the years 1974–1990 edited by Hylleberg (1992).

2.3.7 An important comment

The above preliminary matters are very important and their treatment should not be rushed. Getting a clear time plot and assessing the properties of the data is particularly important. Is trend and/or seasonality present? If so, how should such effects be modelled, measured or removed? In my experience, the treatment of such problem-specific features as calendar effects (e.g. noting whether Easter is in March or April in each year), outliers, possible errors and missing observations can be *more* important than the choice of forecasting method.

Even more basic is to assess the structure and format of the data. Have all important variables been measured? Has the method of measurement changed during the period of observation? Has there been some sort of sudden change during the period of observation? If so, why has it occurred and what should be done about it?

Example 2.1. A seasonal adjustment disaster. While advising a company on ways of improving their seasonal adjustment procedures, I noticed that one group of series had been recorded quarterly until 1995, and monthly thereafter. However, the series had been treated as though they were monthly throughout. Needless to say, the answers that were produced by conventional time-series methods were inappropriate, especially towards the beginning of the series, although it was fortunate that the seasonal adjustments had largely settled down by the time the 1999 data arrived.

It is easy to say that this is just a silly misapplication of time-series methodology, which should not happen in a good statistical department. In fact, many companies have several hundred or even thousands of series to analyse, and this sort of thing can happen all too easily, even in a well-run department. It reminds us that it is very easy to get bogged down in trying to make small improvements to statistical procedures, and lose sight of more important basic questions.   □

2.4 Stationary stochastic processes

This section reviews some basic theory for time series. If future values can be predicted exactly from past values, then a series is said to be *deterministic*. However, most series are *stochastic*, or *random*, in that the future is only partly determined by past values. If an appropriate model for this random behaviour can be found, then the model should enable good forecasts to be computed.

A model for a stochastic time series is often called a *stochastic process*. The latter can be described as a family of random variables indexed by time, and will be denoted by $(X_1, X_2, \ldots)$, or more generally by $\{X_t\}$, in discrete time. An alternative, more precise, notation is to use $\{X_t, t \in T\}$ where T denotes the index set of times on which the process is defined. This notation is necessary when observations are not equally spaced through time, but we restrict attention to the equally spaced case when the index set consisting of the positive integers is commonly used.

We also restrict attention to *real-valued* series which are *causal*. Put simply, the latter restriction means that the process is only allowed to move forward through time so that the value at time t is not allowed to depend on any future values. Of course, if the 'time series' consisted of one-dimensional *spatial* data, such as the heights of a surface at equally spaced points on a horizontal line, then causality is inappropriate.

We regard the observed value at time t, namely x_t, as an observation on an underlying random variable, X_t. The observed time series is called a (finite) *realization* of the stochastic process and the population of all possible realizations is called the *ensemble*. Time-series analysis is different from other statistical problems in that the observed time series is usually the one and only realization that will ever be observed. In other words, x_t is usually the only observation we will ever get on X_t. Despite this, we want to estimate the properties of the underlying stochastic process. Estimation theory for stochastic processes is partially concerned with seeing whether and when a single realization is enough to estimate the properties of the underlying model. Some of these issues are explored in Section 2.6.

At this stage, we concentrate attention on processes which are *stationary*. Put loosely, this means that the properties of the underlying model do not change through time. More formally, a stochastic process is said to be *second-order stationary* if its first and second moments are finite and do not change through time. The first moment is the mean, $E[X_t]$, while the general second moment is the covariance between X_t and X_{t+k} for different values of t and k. This type of covariance is called an *autocovariance*. The variance, $\text{Var}[X_t]$, is a special case of the latter when the lag k is zero. Thus a process is second-order stationary if $E[X_t]$ is a finite constant, say μ, for all t, if $\text{Var}[X_t]$ is a finite constant, say σ^2, for all t, and, more generally, if the autocovariance function depends only on the lag, k, so that

$$\text{Cov}\,[X_t, X_{t+k}] = E[(X_t - \mu)(X_{t+k} - \mu)]$$

$$= \gamma_k \qquad (2.4.1)$$

for all t. The set of autocovariance coefficients $\{\gamma_k\}$, for $k = 0, 1, 2, \ldots$, constitute the *autocovariance function* (abbreviated acv.f.) of the process. Note that γ_0 equals the variance, σ^2. Second-order stationarity is sometimes called *covariance* or *weak* stationarity.

The acv.f. is often standardized to give a set of *autocorrelation coefficients*, $\{\rho_k\}$, given by:

$$\rho_k = \gamma_k / \gamma_0 \qquad (2.4.2)$$

for $k = 0, 1, 2, \ldots$. The $\{\rho_k\}$ constitute the *autocorrelation function* (abbreviated ac.f.) For stationary processes, ρ_k measures the correlation at lag k between X_t and X_{t+k}. The ac.f. is an *even* function of lag, since $\rho_k = \rho_{-k}$, and has the usual property of correlation that $|\rho_k| \leq 1$. Some additional useful functions, which are complementary to the ac.f., include the *partial autocorrelation function* (abbreviated partial ac.f.) which essentially measures the excess correlation at lag k which has not already been accounted for by autocorrelations at lower lags. Another potentially useful tool in model identification is the *inverse* ac.f. (e.g. Wei, 1990, p. 123).

A stochastic process is said to be a *Gaussian* (or normal) process if the joint distribution of any set of X_t's is multivariate normal. Such a process is completely characterized by its first and second moments but it is advisable to remember that this is not so for non-Gaussian processes and that it is possible to find Gaussian and non-Gaussian processes with the same ac.f. This creates obvious difficulties in interpreting sample ac.f.s when trying to identify a suitable underlying model. For non-Gaussian processes it may also be necessary to consider *strict* rather than second-order stationarity, wherein the joint distribution of any set of random variables is not changed by shifting them all by the same time τ, for any value of τ. Although strict stationarity sounds like (and is) a strong condition, note that it does not imply second-order stationarity without the additional assumption that the first and second moments are finite.

An alternative way of describing a stationary stochastic process is by means of its *spectral density function*, or *spectrum*, which is the discrete Fourier transform of $\{\gamma_k\}$, namely:

$$f(\omega) = \frac{1}{\pi} \left(\gamma_0 + 2 \sum_{k=1}^{\infty} \gamma_k \cos \omega k \right) \qquad (2.4.3)$$

for $0 \leq \omega \leq \pi$. The function describes how the overall variance of the process is distributed over different frequencies from zero to the *Nyquist frequency*, π. The latter frequency is the highest frequency about which we can get information from data recorded at unit intervals of time and corresponds to a sine wave which completes one cycle in two time intervals. Note that there are several alternative ways of writing the spectrum in terms of the acv.f. In particular, some authors define $f(\omega)$ as an even function over the frequency range $[-\pi, \pi]$ with a constant $1/2\pi$ outside the bracket in (2.4.3). This definition generalizes more naturally to multidimensional processes but involves the artifact of negative frequencies. A normalized spectrum may also be defined as the discrete Fourier transform of $\{\rho_k\}$.

The two functions $\{\gamma_k\}$ and $f(\omega)$ are equivalent and complementary. An analysis based primarily on estimates of the autocorrelation (or autocovariance) function is called an analysis in the *time domain*. An analysis based primarily on the spectrum is called an analysis in the *frequency domain* or *spectral analysis*. Sometimes the analyst needs to look at both functions, but in time-series forecasting it is rather rare to use frequency domain methods. Thus this book is concerned almost exclusively with the time-domain and the reader is referred to Priestley (1981) and Percival and Walden (1993) for two good references on spectral analysis.

With observations on two or more variables, the cross-correlation function and its Fourier transform, the cross-spectrum, are important tools in describing the underlying process. They are natural bivariate generalizations of the (univariate) autocorrelation function and the (univariate) spectrum. If two stationary processes, say $\{X_t\}$ and $\{Y_t\}$, constitute a bivariate stationary process, then the covariance of X_t and Y_{t+k} will depend only on the lag k and may be written

$$\text{Cov } [X_t, Y_{t+k}] = \text{E} \left[(X_t - \mu_x)(Y_{t+k} - \mu_y) \right] = \gamma_{xy}(k) \qquad (2.4.4)$$

The set of coefficients $\{\gamma_{xy}(k)\}$, for $k = 0, \pm1, \pm2, \ldots$, is called the *cross-covariance function*. The function may be standardized to give the *cross-correlation function*, $\rho_{xy}(k)$, by

$$\rho_{xy}(k) = \gamma_{xy}(k) / \sqrt{\gamma_{xx}(0)\gamma_{yy}(0)} \qquad (2.4.5)$$

Here the denominator is just the product of the standard deviations of the X and Y processes. In many respects, this function has quite different properties from those of the autocorrelation function. Although the values are still restricted to the range $(-1, 1)$, the value at lag zero, namely $\rho_{xy}(0)$, will generally not equal unity (as for the autocorrelation function). Moreover, the cross-correlation function is *not* in general an even function of lag, in that $\rho_{xy}(k)$ may not equal $\rho_{xy}(-k)$ (whereas, for the autocorrelation function we do have $\rho_k = \rho_{-k}$).

The *cross-spectrum* may be defined as

$$f_{xy}(\omega) = \frac{1}{\pi} \left[\sum_{k=-\infty}^{\infty} \gamma_{xy}(k)e^{-i\omega k} \right] \qquad (2.4.6)$$

for $0 \le \omega \le \pi$. This function is generally complex, rather than real, unlike the spectrum of a univariate process which is always real. The interpretation of sample cross-correlations and cross-spectra is rather difficult in practice and discussion is deferred until Chapter 5.

There are many useful classes of model. Some are for a univariate process, while others are multivariate. Some are stationary but others are non-stationary. In the latter case functions like the autocorrelation function will not be meaningful until sources of non-stationarity have been removed. Some classes of univariate model are briefly introduced in Section 2.5 below, and then covered in much greater depth in Chapter 3. Multivariate models are deferred until Chapter 5.

2.5 Some classes of univariate time-series model

Many forecasting procedures are based on a time-series *model*. It is therefore helpful to be familiar with a range of time-series models before starting to look at forecasting methods. This section aims to give an introductory flavour of some simple important univariate models. A *univariate* model describes the distribution of a single random variable at time t, namely X_t, in terms of its relationship with past values of X_t and its relationship with a series of white-noise random shocks as defined in Section 2.5.1 below.[6]

2.5.1 *The purely random process*

The simplest type of model, used as a 'building brick' in many other models, is the *purely random process*. This process may be defined as a sequence of uncorrelated, identically distributed random variables with zero mean and constant variance. This process is clearly stationary and has ac.f. given by

$$\rho_k = \begin{cases} 1 & k = 0 \\ 0 & \text{otherwise} \end{cases} \qquad (2.5.1)$$

The spectrum of this process is constant (as would intuitively be expected).

This process is variously called (uncorrelated) *white noise*, the *innovations* process or (loosely) the 'error' process. The model is rarely used to describe data directly, but is often used to model the random disturbances in a more complicated process. If this is done, the white-noise assumptions need to be checked.

Some authors use the term 'purely random process' to refer to a sequence of *independent*, rather than uncorrelated, random variables. There is no difference in regard to second-order properties and of course independence implies lack of correlation. Moreover, the converse is true when the process is a *Gaussian* process (as it will often be assumed to be). However, for non-linear models the difference between uncorrelated white noise and independent white noise may be crucial, and the stronger assumption of independence is generally needed when looking at non-linear models and predictors. In future, we will highlight this difference where necessary by describing a purely random process consisting of independent random variables as being *strict* or *independent white noise*.

By convention, we use $\{Z_t\}$ in this book to denote a purely random process with zero mean and variance σ_z^2 when it is a component of a random walk model, of an autoregressive model or of the more general class of (Box-Jenkins) ARIMA processes, though for other classes of model we typically use the alternative notation $\{\varepsilon_t\}$. Other writers use a variety of symbols for the same process, including $\{a_t\}$ and $\{u_t\}$. Some writers always assume serial independence, rather than a lack of serial correlation, but, as noted above, for a linear Gaussian process this makes no difference. We generally make the minimal assumption of zero correlation in the linear case but assume independence in the non-linear case.

---

[6] Some writers also allow a deterministic function of time to be included, but there are definitely no terms involving any other explanatory variables.

2.5.2 The random walk

A model of much practical interest is the *random walk* which is given by

$$X_t = X_{t-1} + Z_t \qquad (2.5.2)$$

where $\{Z_t\}$ denotes a purely random process. This model may be used, at least as a first approximation, for many time series arising in economics and finance (e.g. Meese and Rogoff, 1983). For example, the price of a particular share on a particular day is equal to the price on the previous trading day plus or minus the change in share price. It turns out that the latter quantity is generally not forecastable and has properties similar to those of the purely random process.

The series of random variables defined by (2.5.2) does *not* form a stationary process as it is easy to show that the variance increases through time. However, the first differences of the series, namely $(X_t - X_{t-1})$, *do* form a stationary series. We will see that taking differences is a common procedure for transforming a non-stationary series into a stationary one.

2.5.3 Autoregressive processes

A process $\{X_t\}$ is said to be an *autoregressive* process of order p (abbreviated AR(p)) if

$$X_t = \phi_1 X_{t-1} + \phi_2 X_{t-2} + \cdots + \phi_p X_{t-p} + Z_t \qquad (2.5.3)$$

Thus the value at time t depends linearly on the last p values and the model looks like a regression model – hence the term *auto*regression.

The simplest example of an AR process is the first-order case, denoted AR(1), given by

$$X_t = \phi X_{t-1} + Z_t \qquad (2.5.4)$$

Clearly, if $\phi = 1$, then the model reduces to a random walk as in (2.5.2), when the model is non-stationary. If $|\phi| > 1$, then it is intuitively obvious that the series will be explosive and hence non-stationary. However, if $|\phi| < 1$, then it can be shown that the process is stationary,[7] with ac.f. given by $\rho_k = \phi^k$ for $k = 0, 1, 2, \ldots$. Thus the ac.f. decreases exponentially. Further details about AR models are given in Section 3.1.1.

2.5.4 Moving average processes

A process $\{X_t\}$ is said to be a moving average process of order q (abbreviated MA(q)) if

$$X_t = Z_t + \theta_1 Z_{t-1} + \cdots + \theta_q Z_{t-q} \qquad (2.5.5)$$

Thus the value at time t is a sort of moving average of the (unobservable) random shocks, $\{Z_t\}$. However, the 'weights', $\{\theta_j\}$, involved in the moving

---

[7] Section 3.1.4 discusses how to make this statement more rigorous.

average will generally not add to unity and so the phrase 'moving average' is arguably unhelpful.

The simplest example of an MA process is the first-order case, denoted MA(1), given by

$$X_t = Z_t + \theta Z_{t-1} \qquad (2.5.6)$$

It can be shown that this process is stationary for all values of θ with an ac.f. given by

$$\rho_k = \begin{cases} 1 & k = 0 \\ \theta/(1+\theta^2) & k = 1 \\ 0 & k > 1 \end{cases}$$

Thus the ac.f. 'cuts off' at lag 1. Further details about MA processes are given in Section 3.1.2.

2.5.5 The random walk plus noise model

The *random walk plus noise* model, sometimes called the *local level,* or *steady* model is a simple example of a class of models called *state-space* models. The latter are considered in more detail in Section 3.2. Suppose the observed random variable at time t may be written in the form

$$X_t = \mu_t + n_t \qquad (2.5.7)$$

where the local level, μ_t, changes through time like a random walk so that

$$\mu_t = \mu_{t-1} + w_t \qquad (2.5.8)$$

The two sources of random variation in the above equations, namely n_t and w_t, are assumed to be independent white noise processes with zero means and respective variances σ_n^2, σ_w^2. The properties of this model depend on the ratio of the two error variances, namely σ_w^2/σ_n^2, which is called the *signal-to-noise ratio*. In the jargon of state-space modelling, the unobserved variable, μ_t, which denotes the local level at time t, is called a *state variable*, (2.5.7) is called the *observation* or *measurement* equation, while (2.5.8) is called the *transition* equation.[8]

2.6 The correlogram

The correlogram is probably the most useful tool in time-series analysis after the time plot. It can be used at two different levels of sophistication, either as a relatively simple descriptive tool or as part of a more general procedure for identifying an appropriate model for a given time series. We say more about the latter in Chapter 3 and concentrate here on simpler

---

[8] Note that various alternative representations of this model are in common use. For example, some authors write the transition equation (2.5.8) with μ_{t+1} on the left-hand side and μ_t and w_t on the right-hand side so that the times match up on the right-hand side. The reader needs to ensure that any representation makes sense and gives internally consistent results.

ideas, though this section does provide the opportunity to comment on some basic estimation issues in regard to time-series modelling.

Denote the observed time series by $\{x_1, x_2, \ldots, x_N\}$. The sample autocovariance coefficient at lag k is usually calculated by

$$c_k = \sum_{t=1}^{N-k} (x_t - \bar{x})(x_{t+k} - \bar{x})/N \qquad (2.6.1)$$

for $k = 0, 1, 2, \ldots$, and the sample autocorrelation coefficient at lag k is then calculated by

$$r_k = c_k/c_0 \qquad (2.6.2)$$

The graph of r_k against k is called the *sample autocorrelation function* (abbreviated *ac.f.*) or the *correlogram*. It is an important tool in assessing the behaviour and properties of a time series. It is typically plotted for the original series and also after differencing or transforming the data as necessary to make the series look stationary and approximately normally distributed.

For data from a stationary process, it can be shown that the correlogram generally provides an estimate of the theoretical ac.f. defined in (2.4.2). Although intuitively 'obvious', this is mathematically hard to prove because it requires that averages over time for an observed time series (like $\bar{x}$) enable us to estimate the ensemble properties of the underlying process (like $E(X_t)$); in other words, that we can estimate the properties of the random variable at time t with the help of observations made at other times.[9] It follows that, for data from a non-stationary process, the correlogram does not provide an estimate of anything! In that case, the values in the correlogram will typically not come down to zero except at high lags, and the only merit of the correlogram is to indicate that the series is not stationary.

Interpreting a correlogram is one of the hardest tasks in time-series analysis, especially when N is less than about 100 so that the sample autocorrelations have relatively large variance. For a stationary series, the pattern of the correlogram may suggest a a stationary model with an ac.f. of similar shape. The simplest case is that of a purely random process, where it can be shown that r_k is asymptotically normally distributed with mean $-1/N$, and standard deviation $1/\sqrt{N}$ for $k \neq 0$. As the mean, $-1/N$, is small compared with the standard deviation, it is customary to take the mean as being approximately zero and regard values outside the range $0 \pm 2/\sqrt{N}$ as being significantly different from zero. Several significant

---

[9] Strictly speaking, the process needs to have appropriate 'ergodic' properties so that averages over time from a single realization provide estimates of the properties of the underlying process (e.g. Priestley, 1981, Section 5.3.6). For example a covariance-stationary process is said to be 'ergodic in the mean' if the sample mean, $\bar{x} = \sum_{t=1}^{N} x_t$, converges in probability to $E(X_t)$ as $N \to \infty$ so that $\bar{x}$ provides a consistent estimate of the ensemble average. A sufficient condition is that $\rho_k \to 0$ as $k \to \infty$. We will assume appropriate ergodic properties are satisfied throughout the book.

coefficients, especially at important low lags, provide strong evidence that the data do *not* come from a purely random process.

One common pattern for stationary time series is to see *short-term correlation* typified by finding perhaps the first three or four values of r_k to be significantly different from zero. If they seem to decrease in an approximately exponential way then an AR(1) model is indicated. If they behave in a more complicated way, then a higher-order AR model may be appropriate. If the only significant autocorrelation is at lag one, then a MA(1) model is indicated. For seasonal series, there is likely to be a large positive value of r_k at the seasonal period, and this may still be present to a (much) lesser extent even after the seasonal effect has supposedly been removed. Thus the correlogram is often used to see if seasonality is present. For series with a trend, the correlogram will not come down to zero until a high lag has been reached, perhaps up towards half the length of the series. The correlogram provides little information in the presence of trend other than as an indicator that some form of trend-removal is necessary to make the series stationary.

As with other material in this brief introduction to time-series analysis, the reader is referred to the many introductory texts on time series for a more complete guide to the interpretation of the correlogram. In addition, the novice is advised to get experience of looking at correlograms, both for simulated data from a known model and for real data. As is often the case, practical experience is the best teacher.

Example 2.2.   Correlograms for the S&P 500 series. We give one simple example to indicate how to interpret correlograms. The S&P 500 index series is plotted in Figure 1.1. The plot shows no evidence of seasonality and no systematic trend, although the local mean does seem to change (both up and down). The more experienced reader will realize that the changes in local mean indicate that the series is not stationary, and the correlogram of the series, shown in Figure 2.2(a), confirms this. The autocorrelations up to lag 13 are all positive, after which there is a sequence of negative autocorrelations from lag 14 to a lag in excess of 20 (actually until lag 37). This is typical behaviour for a non-stationary series. The dotted lines on the graph, at around 0.21, indicate that values outside these dotted lines are significantly different from zero. In this case, there are so many 'significant' values that modelling is clearly problematic. The series must be made more stationary (whatever that means) before we can usefully interpret the correlogram.

If there were an obvious linear trend in the mean, we could fit a straight line to the data and look at the deviations from the line. This is not appropriate here. Instead we form the series of first differences and compute its correlogram. This is shown in Figure 2.2(b). Apart from the value at lag zero (which is always one and tells us nothing), the autocorrelations all lie inside the dotted lines. There is effectively little or no autocorrelation left in the series. In other words, the first differences of the S&P 500 series are very close to being a completely random series. Put another way, this means that the original series is very close to being a random walk (see Section 2.5.2).

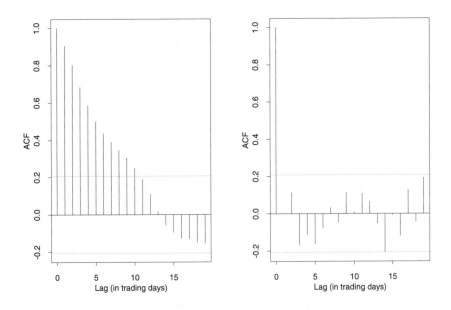

Figure 2.2. *The sample correlogram of (a) the S&P 500 index series plotted in Figure 1.1; (b) The first differences of the S&P 500 series.*

The econometrician might prefer to induce stationarity by looking at the series of percentage changes, namely $100 \times (I_t - I_{t-1})/I_{t-1}$, where I_t denotes the value of the index at time t. An alternative is to work with the series of so-called (compounded) returns given by $100 \times \log(I_t/I_{t-1})$. The first series involves the first differences, $(I_t - I_{t-1})$, while the second involves the first differences of the logs, namely $(\log I_t - \log I_{t-1})$. For the relatively short series considered here, where I_t does not vary much in size, it makes little difference which series is used. For example the correlogram of the first differences of the logarithms is virtually identical to that in Figure 2.2(b). For longer financial series, some form of percentage variation will normally be preferred to first differences.   □

With observations on two or more series, it is possible to look, not only at the correlogram of each series, but also to look at the correlations *between* the series. The *sample cross-correlation function* is defined in Chapter 5 and discussion is deferred until then.

CHAPTER 3

Univariate Time-Series Modelling

Many univariate forecasting procedures are based explicitly or implicitly on a univariate time-series *model*. It is therefore necessary for the forecaster to be familiar with a range of possible models and have a thorough understanding of the modelling process as applied to time-series data.

A few simple models were briefly introduced in Section 2.5 and some general hints on model building were made throughout Chapter 2, especially in regard to using the time plot and the correlogram. This chapter discusses a wider variety of models in greater depth and then makes some general comments on models and the model-building process. Some topics will be familiar to some readers, but the chapter also introduces some important topics which are likely to be new to many readers, including long-memory models, some non-linear models, and some aspects of model formulation and model selection.

3.1 ARIMA models and related topics

The ARIMA class of models is an important forecasting tool, and is the basis of many fundamental ideas in time-series analysis. The acronym ARIMA stands for 'autoregressive integrated moving average', and the different components of this general class of models will be introduced in turn. The original key reference[1] is Box and Jenkins (1970), and ARIMA models are sometimes called *Box-Jenkins models*. Some closely related models will also be discussed.

We concentrate here on the theoretical properties of the models. Some general comments on modelling are given later in Section 3.5, while some specific comments on fitting ARIMA models are given in Section 4.2. Many more detailed accounts are given elsewhere in the literature, but, with so much good software now available to fit ARIMA models, it is no longer necessary for the user to have detailed knowledge of what is involved (though, as always, it is a good idea to know what computer software is doing in broad terms so as to avoid inappropriate software use).

3.1.1 Autoregressive (AR) processes

A time series $\{X_t\}$ is said to be an autoregressive process of order p (abbreviated AR(p)) if it is a weighted linear sum of the past p values

---

[1] Now revised as Box et al. (1994).

plus a random shock so that

$$X_t = \phi_1 X_{t-1} + \phi_2 X_{t-2} + \cdots + \phi_p X_{t-p} + Z_t \qquad (3.1.1)$$

where $\{Z_t\}$ denotes a purely random process with zero mean and variance σ_z^2. Using the backward shift[2] operator B, such that $BX_t = X_{t-1}$, the AR(p) model may be written more succinctly in the form

$$\phi(B)X_t = Z_t \qquad (3.1.2)$$

where $\phi(B) = 1 - \phi_1 B - \phi_2 B^2 - \cdots - \phi_p B^p$ is a polynomial in B of order p. The properties of AR processes defined by (3.1.1) can be examined by looking at the properties of the function ϕ. As B is an operator, the algebraic properties of ϕ have to be investigated by examining the properties of $\phi(x)$, say, where x denotes a complex variable, rather than by looking at $\phi(B)$. It can be shown that (3.1.2) has a unique causal stationary solution provided that the roots of $\phi(x) = 0$ lie outside the unit circle. This solution may be expressed in the form

$$X_t = \sum_{j \geq 0}^{\infty} \psi_j Z_{t-j} \qquad (3.1.3)$$

for some constants ψ_j such that $\sum |\psi_j| < \infty$.

The above statement about the unique stationary solution of (3.1.2) may be unfamiliar to the reader who is used to the more customary time-series literature. The latter typically says something like "an AR process is stationary provided that the roots of $\phi(x) = 0$ lie outside the unit circle". This will be good enough for most practical purposes but is not strictly accurate; for further remarks on this point, see Section 3.1.4 and Brockwell and Davis (1991).

The simplest example of an AR process is the first-order case given by

$$X_t = \phi X_{t-1} + Z_t \qquad (3.1.4)$$

The time-series literature typically says that an AR(1) process is stationary provided that $|\phi| < 1$. It is more accurate to say that there is a unique stationary solution of (3.1.4) which is causal, provided that $|\phi| < 1$. The ac.f. of a stationary AR(1) process[3] is given by $\rho_k = \phi^k$ for $k = 0, 1, 2, \ldots$. For higher-order stationary AR processes, the ac.f. will typically be a mixture of terms which decrease exponentially or of damped sine or cosine waves. The ac.f. can be found by solving a set of difference equations called the Yule-Walker equations given by

$$\rho_k = \phi_1 \rho_{k-1} + \phi_2 \rho_{k-2} + \cdots + \phi_p \rho_{k-p} \qquad (3.1.5)$$

---

[2] Note that economists typically use the notation L (for lag operator) rather than B (for backward shift operator).

[3] Strictly speaking we should say that it is the ac.f. of the unique stationary solution of (3.1.4) – see the later theoretical subsection. In future we will not continue to make these rather pedantic points but revert throughout to the usual time-series usage of the phrase 'stationary AR process' as meaning the unique causal stationary solution of an equation of the form (3.1.1).

for $k = 1, 2, \ldots$, where $\rho_0 = 0$. Notice that the AR model is typically written in *mean-corrected* form with no constant on the right-hand side of (3.1.1). This makes the mathematics much easier to handle.

A useful property of an AR(p) process is that it can be shown that the partial ac.f. is zero at all lags greater than p. This means that the sample partial ac.f. can be used to help determine the order of an AR process (assuming the order is unknown as is usually the case) by looking for the lag value at which the sample partial ac.f. 'cuts off' (meaning that it should be approximately zero, or at least not significantly different from zero, for higher lags).

3.1.2 Moving average (MA) processes

A time series $\{X_t\}$ is said to be a moving average process of order q (abbreviated MA(q)) if it is a weighted linear sum of the last q random shocks so that

$$X_t = Z_t + \theta_1 Z_{t-1} + \cdots + \theta_q Z_{t-q} \qquad (3.1.6)$$

where $\{Z_t\}$ denotes a purely random process with zero mean and constant variance σ_z^2. (3.1.6) may alternatively be written in the form

$$X_t = \theta(B) Z_t \qquad (3.1.7)$$

where $\theta(B) = 1 + \theta_1 B + \cdots + \theta_q B^q$ is a polynomial in B of order q. Note that some authors (including Box et al., 1994) parameterize an MA process by replacing the plus signs in (3.1.6) with minus signs, presumably so that $\theta(B)$ has a similar form to $\phi(B)$ for AR processes, but this seems less natural in regard to MA processes. There is no difference in principle between the two notations but the signs of the θ values are reversed and this can cause confusion when comparing formulae from different sources or examining computer output.

It can be shown that a finite-order MA process is stationary for all parameter values. However, it is customary to impose a condition on the parameter values of an MA model, called the *invertibility*[4] *condition*, in order to ensure that there is a unique MA model for a given ac.f. This condition can be explained as follows. Suppose that $\{Z_t\}$ and $\{Z_t'\}$ are independent purely random processes and that $\theta \in (-1, 1)$. Then it is straightforward to show that the two MA(1) processes defined by $X_t = Z_t + \theta Z_{t-1}$ and $X_t = Z_t' + \theta^{-1} Z_{t-1}'$ have exactly the same autocorrelation function (ac.f.). Thus the polynomial $\theta(B)$ is not uniquely determined by the ac.f. As a consequence, given a sample ac.f., it is not possible to estimate a unique MA process from a given set of data without putting some constraint on what is allowed. To resolve this ambiguity, it is usually required that the polynomial $\theta(x)$ has all its roots outside the unit circle.

---

[4] Control engineers use the term 'minimum phase' rather than 'invertible'.

It then follows that we can rewrite (3.1.6) in the form

$$X_t - \sum_{j \geq 1} \pi_j X_{t-j} = Z_t \qquad (3.1.8)$$

for some constants π_j such that $\sum |\pi_j| < \infty$. In other words, we can *invert* the function taking the Z_t sequence to the X_t sequence and recover Z_t from present and past values of X_t by a convergent sum. The negative sign of the π-coefficients in (3.1.8) is adopted by convention so that we are effectively rewriting an MA process of finite order as an AR(∞) process. The astute reader will notice that the invertibility condition (roots of $\theta(x)$ lie outside unit circle) is the mirror image of the condition for stationarity of an AR process (roots of $\phi(x)$ lie outside the unit circle).

The ac.f. of an MA(q) process can readily be shown to be

$$\rho_k = \begin{cases} 1 & k = 0 \\ \sum_{i=0}^{q-k} \theta_i \theta_{i+k} / \sum_{i=0}^{q} \theta_i^2 & k = 1, 2, \ldots, q \\ 0 & k > q \end{cases} \qquad (3.1.9)$$

where $\theta_0 = 1$. Thus the ac.f. 'cuts off' at lag q. This property may be used to try to assess the *order* of the process (i.e. What is the value of q?) by looking for the lag beyond which the sample ac.f. is not significantly different from zero.

The MA process is relevant to a mathematical theorem called the *Wold decomposition theorem* – see, for example, Priestley, 1981, Section 10.1.5. In brief simplified form, this says that any stationary process can be expressed as the sum of two types of processes, one of which is *non-deterministic* while the other is (linearly) *deterministic*. These terms are defined as follows. If the process can be forecast exactly by a linear regression on past values, even if recent values are not available, then the process is called *deterministic* (or singular). However, if linear regression on the remote past is useless for prediction purposes, then the process is said to be *non-deterministic* (or purely indeterministic or regular or stochastic).

The connection with MA processes is as follows. It can be shown that the non-deterministic part of the Wold decomposition can be expressed as an MA process of possibly *infinite* order with the requirement that successive values of the Z_t sequence are uncorrelated rather than independent as assumed by some authors when defining MA processes. Formally, any stationary nondeterministic time series can be expressed in the form

$$X_t = \sum_{j=0}^{\infty} \psi_j Z_{t-j} \qquad (3.1.10)$$

where $\psi_0 = 1$ and $\sum_{j=0}^{\infty} \psi_j^2 < \infty$, and $\{Z_t\}$ denotes a purely random process (or uncorrelated white noise) with zero mean and constant variance, σ^2, which is uncorrelated with the deterministic part of the process (if any). The $\{Z_t\}$ are sometimes called *innovations*, as they are the one-step-ahead forecast errors when the best linear predictor is used to make one-step-ahead forecasts. The formula in (3.1.10) is an MA(∞) process, which is often called the *Wold representation* of the process. It is also sometimes called a *linear process* or, as in this book, a *general linear process*. However, note that the latter terms are used by some writers when the Z_t's are independent, rather than uncorrelated, or when the summation in (3.1.10) is from $-\infty$ to $+\infty$, rather than 0 to ∞. Of course, if the Z_t's are normally distributed, then zero correlation implies independence anyway and we have what is sometimes called a *Gaussian linear process*.

In practice the Wold decomposition is of rather limited value. If the generating sequence of the process is non-linear, the Wold decomposition is generally unhelpful as the best mean square predictor may be quite different from the best linear predictor. Moreover, even if the process is linear, the MA(∞) representation of a stochastic process involves an infinite number of parameters which are impossible to estimate from a finite set of data. Thus it is customary to search for a model that is a parsimonious approximation to the data, by which is meant using as few parameters as possible. One common way to proceed is to consider the class of mixed ARMA processes as described below.

3.1.3 ARMA processes

A mixed autoregressive moving average model with p autoregressive terms and q moving average terms is abbreviated ARMA(p, q) and may be written as

$$\phi(B)X_t = \theta(B)Z_t \qquad (3.1.11)$$

where $\phi(B), \theta(B)$ are polynomials in B of finite order p, q, respectively. This combines (3.1.2) and (3.1.7). Equation 3.1.11 has a unique causal stationary solution provided that the roots of $\phi(x) = 0$ lie outside the unit circle. The process is invertible provided that the roots of $\theta(x) = 0$ lie outside the unit circle. In the stationary case, the ac.f. will generally be a mixture of damped exponentials or sinusoids.

The importance of ARMA processes is that many real data sets may be approximated in a more parsimonious way (meaning fewer parameters are needed) by a mixed ARMA model rather than by a pure AR or pure MA process. We know from the Wold representation in (3.1.10) that any stationary process can be represented as a MA(∞) model, but this may involve an infinite number of parameters and so does not help much in modelling. The ARMA model can be seen as a model in its own right or as an approximation to the Wold representation. In the latter case, the generating polynomial in B, which gives (3.1.10), namely

$$\psi(B) = \sum_{j=0}^{\infty} \psi_j B^j$$

may be of infinite order, and so we try to approximate it by a rational polynomial of the form

$$\psi(B) = \theta(B)/\phi(B)$$

which effectively gives the ARMA model.

3.1.4 Some theoretical remarks on ARMA processes

This subsection[5] takes a more thorough look at some theoretical aspects of ARMA processes (and hence of pure AR and MA processes by putting q or p equal to one). The general ARMA(p,q) process is defined by (3.1.11). Time-series analysts typically say that this equation *is* an ARMA(p,q) process. However, strictly speaking, the above equation is just that – an equation and not a process. In contrast, the definition of a process should uniquely define the sequence of random variables $\{X_t\}$. Like other difference equations, (3.1.11) will have infinitely many solutions (except for pure MA processes) and so, although it is possible to say that any solution of (3.1.11) is an ARMA(p,q) process, this does not of itself uniquely define the process.

Consider, for example, the AR(1) process, defined earlier in (3.1.4), for which

$$X_t = \phi X_{t-1} + Z_t \qquad\qquad (3.1.12)$$

for $t = 0, \pm 1, \pm 2, \ldots$. As noted earlier, it is customary for the time-series literature to say that an AR(1) process is stationary provided that $|\phi| < 1$. However, the first-order difference equation defined by (3.1.12) has infinitely many solutions. For example, the reader may check that the following process is a solution to (3.1.12):

$$X_t = \sum_{r \geq 0} \phi^r Z_{t-r} + \phi^t K \qquad\qquad (3.1.13)$$

where K denotes a constant. If we take K to be zero, then we obtain the unique stationary solution, but for any other value of K, the process will not be stationary. However, it is readily apparent that the general solution will tend to the stationary solution as $t \to \infty$. In this regard the question of whether or not the equation is *stable* becomes important. The property of *stability* is linked to stationarity. If we regard (3.1.12) as a linear filter for changing an input process $\{Z_t\}$ to an output process $\{X_t\}$, then it can be shown that the system is stable provided that $|\phi| < 1$. This means that the effect of any perturbation to the input will eventually die away. This can be used to demonstrate that any deviation from the stationary solution will also die away. Thus the general solution tends to the stationary solution as t increases.

From both a practical and theoretical point of view, it is usually advisable to restrict attention to the unique stationary solution of an ARMA equation. This book follows current time-series practice in using the phrase 'stationary ARMA process' to mean the unique causal stationary solution of an ARMA equation. Some writers (e.g. Brockwell and Davis, 1996) avoid this difficulty by defining a process $\{X_t\}$ to be an ARMA process only if it is stationary and hence have no need to add the adjective 'stationary'.

The reader will recall that we have restricted attention to so-called *causal*

---

[5] This subsection may be omitted by the reader with little interest in rigorous theory.

processes, where the value of an observed time series is only allowed to depend on present and past values of other variables. In other words, time is only allowed to move in one direction, namely forwards. If we allow ϕ to exceed unity for an AR(1) process, then it is readily seen that the series defined by (3.1.12) will typically 'explode'. In fact, it *is* possible to find a stationary sequence which satisfies (3.1.12), with $|\phi| > 1$, by allowing X_t to depend on *future* values of Z_t. However, for time-series data, this is usually regarded as unnatural. In any case it can be shown that any stationary AR(1) process, with $|\phi| > 1$, can be re-expressed as an AR(1) process, with $|\phi| < 1$ and the same second-order properties, but based on a different white noise sequence. Thus nothing is lost by restricting attention to the case $|\phi| < 1$. Note that if $\phi = 1$, then there is no stationary solution.

A stationary ARMA process may be written as an MA(∞) process by rewriting (3.1.11) in the form

$$X_t = [\theta(B)/\phi(B)]Z_t$$

and expanding θ/ϕ as a power series in B. For a causal process, the resulting representation may be written in the same form as the Wold representation in (3.1.10), or (3.1.3), namely

$$X_t = \sum_{j=0}^{\infty} \psi_j Z_{t-j} \tag{3.1.14}$$

or as

$$X_t = \psi(B)Z_t \tag{3.1.15}$$

where $\psi(B) = \psi_0 + \psi_1 B + \psi_2 B^2 + \ldots$. A finite MA process is always stationary as there is a finite sum of Z's. However, for an infinite sequence, such as that in (3.1.14), the weighted sum of Z's does not necessarily converge. From (3.1.14), it can readily be shown that

$$\text{Variance}(X_t) = \sigma_Z^2 \sum \psi_j^2 \tag{3.1.16}$$

and so we clearly require $\sum \psi_j^2 < \infty$ for the variance to be finite. In fact, we really need the stronger condition that $\sum |\psi_j| < \infty$ in order to ensure that the system is stable and that necessary sums ensure converge (see Box et al., 1994; Brockwell and Davis, 1991). The latter condition (which is sufficient rather than necessary) is equivalent to the requirement for (causal) stationarity that the roots of $\phi(x)$ lie outside the unit circle.

It may not be immediately obvious why it may be helpful to re-express the ARMA process in (3.1.11) as an MA(∞) process in the form (3.1.14). In fact, the MA(∞) representation is generally the easiest way to find the variance of forecast errors – see Chapter 7. For computing point forecasts, it may also be helpful to try to re-express the process as an AR(∞) process. It turns out that, if the process is invertible, then it is possible to rewrite (3.1.11) in the form of (3.1.8) as

$$\pi(B)X_t = Z_t \tag{3.1.17}$$

where, by convention, we take $\pi(B) = 1 - \sum_{j=1}^{\infty} \pi_j B^j$, and where $\sum |\pi_j| < \infty$ so that the π's are summable.

The ARMA process in (3.1.11) has $\phi(B)$ and $\theta(B)$ of finite order, whereas when a mixed model is expressed in pure MA or AR form, the polynomials $\pi(B)$ and $\psi(B)$ will be of infinite order. We have seen that an ARMA process is stationary if the roots of $\phi(x) = 0$ lie outside the unit circle and invertible if the roots of $\theta(x)$

lie outside the unit circle. We can find corresponding conditions in terms of $\pi(B)$ or $\psi(B)$. It can be shown that the process is stationary if $\psi(x)$ converges on, and within, the unit circle, and invertible if $\pi(x)$ converges on and within the unit circle. The connection between the corresponding conditions here may not be immediately obvious but arises in the following way. Suppose, for example, that the roots of $\phi(x)$ all lie *outside* the unit circle so that the process is stationary. Then there are no roots *on or inside* the unit circle and so $1/\phi(x)$ cannot become infinite on or within the unit circle. A similar argument applies to the invertibility condition.

Finally, we comment that whether or not a process is stationary is unrelated to whether or not it is invertible. It is possible for any combination of the two properties to occur. Note that a pure (i.e. finite-order) AR process is always invertible, even if it is non-stationary, while a pure MA process is always stationary, even if it is not invertible.

Example 3.1   For a mixed model, it is usually easiest to check whether the process is (a) stationary and (b) invertible, by looking to see if the roots of $\phi(x)$ and of $\theta(x)$ lie outside the unit circle, rather than looking at whether $\psi(x)$ and $\pi(x)$ converge on, and within, the unit circle. For example, consider the ARMA$(1,1)$ process

$$(1 - 1.3B)X_t = (1 - 0.5B)Z_t \qquad (3.1.18)$$

The root of $\phi(x) = 0$ is $x = 1/1.3$ which lies inside the unit circle. Thus the process is non-stationary. If we try to write the process as an MA(∞) process, we would find that the coefficients of B^j get larger and larger in amplitude and so the process does not converge on the unit circle. However, note that by equating coefficients of B^j in the equation $\theta(B) = \phi(B)\,\psi(B)$, we can evaluate the ψ_j coefficients for any finite value of j and this can be helpful in assessing the forecasts from a non-stationary model, for example, to evaluate the forecast error variance.

Returning to (3.1.18), the root of $\theta(B) = 0$ is $x = 1/0.5$ which does lie outside the unit circle. Thus the process is invertible. This means that the sequence of π-weights will converge on and within the unit circle. We find $\pi(B) = \phi(B)/\theta(B) = (1 - 1.3B)/(1 - 0.5B) = (1 - 1.3B)(1 + 0.5B + 0.5^2B^2 + \ldots)$, from which we see that π_j, the coefficient of B^j is $0.5^j - 1.3 \times 0.5^{j-1} = (-0.80) \times 0.5^{j-1}$ which clearly gets smaller as j increases.   □

3.1.5 ARIMA processes

We now reach the more general class of models which is the title of the whole of this section. In practice many (most?) time series are non-stationary and so we cannot apply stationary AR, MA or ARMA processes directly. One possible way of handling non-stationary series is to apply *differencing* so as to make them stationary. The first differences, namely $(X_t - X_{t-1}) = (1 - B)X_t$, may themselves be differenced to give second differences, and so on. The dth differences may be written as $(1 - B)^d X_t$. If the original data series is differenced d times before fitting an ARMA(p, q) process, then the model for the original undifferenced series is said to be an ARIMA(p, d, q) process where the letter 'I' in the acronym stands for *integrated* and d denotes the number of differences taken.

Mathematically, (3.1.11) is generalized to give:

$$\phi(B)(1 - B)^d X_t = \theta(B)Z_t \tag{3.1.19}$$

The combined AR operator is now $\phi(B)(1-B)^d$. If we replace the operator B in this expression with a variable x, it can be seen straight away that the function $\phi(x)(1 - x)^d$ has d roots on the unit circle (as $(1 - x) = 0$ when $x = 1$) indicating that the process is non-stationary – which is why differencing is needed, of course!

Note that when $\phi(B)$ and $\theta(B)$ are both just equal to unity (so that p and q are both zero) and d equals one, then the model reduces to an ARIMA$(0, 1, 0)$ model, given by

$$X_t - X_{t-1} = Z_t \tag{3.1.20}$$

This is obviously the same as the *random walk* model in (2.5.2) which can therefore be regarded as an ARIMA$(0, 1, 0)$ model.

When fitting AR and MA models, the main difficulty is assessing the order of the process (i.e. what is p and q?) rather than estimating the coefficients (the ϕ's and θ's). With ARIMA models, there is an additional problem in choosing the required order of differencing (i.e. what is d?). Some formal procedures are available, including testing for the presence of a unit root (see Section 3.1.9), but many analysts simply difference the series until the correlogram comes down to zero 'fairly quickly'. First-order differencing is usually adequate for non-seasonal series, though second-order differencing is occasionally needed. Once the series has been made stationary, an ARMA model can be fitted to the differenced data in the usual way.

3.1.6 SARIMA processes

If the series is *seasonal*, with s time periods per year, then a seasonal ARIMA (abbreviated SARIMA) model may be obtained as a generalization of (3.1.19). Let B^s denote the operator such that $B^s X_t = X_{t-s}$. Thus seasonal differencing may be written as $(X_t - X_{t-s}) = (1 - B^s)X_t$. A seasonal autoregressive term, for example, is one where X_t depends linearly on X_{t-s}. A SARIMA model with non-seasonal terms of order (p, d, q) and seasonal terms of order (P, D, Q) is abbreviated a SARIMA$(p, d, q) \times (P, D, Q)_s$ model and may be written

$$\phi(B)\Phi(B^s)(1 - B)^d(1 - B^s)^D X_t = \theta(B)\Theta(B^s)Z_t \tag{3.1.21}$$

where Φ, Θ denote polynomials in B^s of order P, Q, respectively.

One model, which is particularly useful for seasonal data, is the SARIMA model of order $(0, 1, 1) \times (0, 1, 1)_s$. For monthly data, with $s = 12$, the latter may be written

$$(1 - B)(1 - B^{12})X_t = (1 + \theta B)(1 + \Theta B^{12})Z_t \tag{3.1.22}$$

This model is often called the *airline* model because it was used by Box et al. (1994) to model the logarithms[6] of the airline data, which is plotted in Figure 2.1. Although a series modelled by (3.1.22) is non-stationary, it can readily be shown that the differenced series (after one seasonal and one non-seasonal difference) *is* stationary with 'significant' autocorrelations at lags 1 and 12 only – see Example 3.3 at the end of this chapter.

When fitting SARIMA models, the analyst must first choose suitable values for the two orders of differencing, both seasonal (D) and non-seasonal (d), so as to make the series stationary and remove (most of) the seasonality. Then an ARMA-type model is fitted to the differenced series with the added complication that there may be AR and MA terms at lags which are a multiple of the season length s. The reader is referred to Box et al. (1994, Chapter 9) for further details.

3.1.7 Periodic AR models

If seasonal variation exists through the year, there is no particular reason why model coefficients should stay constant throughout the year. *Periodic Autoregressive* (abbreviated PAR) models provide a variant to SARIMA models wherein the values of the autoregressive parameters are allowed to vary through the seasonal cycle. More generally *periodic correlation* arises when the size of autocorrelation coefficients depends, not only on the lag, but also on the position in the seasonal cycle.

The above ideas will be illustrated with the simplest possible PAR model, namely a stationary zero-mean first-order PAR model, denoted PAR(1). The ordinary AR(1) model has a single AR parameter, say ϕ_1, where $X_t = \phi_1 X_{t-1} + Z_t$. For a PAR(1) model, the autoregressive parameter at lag one, ϕ_1, depends on the position within the seasonal cycle. Thus for quarterly data, there will be four first-order parameters, say $\phi_{1,1}, \phi_{1,2}, \phi_{1,3}, \phi_{1,4}$ corresponding to the four quarters. In order to write down an equation for the model, we need a new notation which indicates time and the position within the seasonal cycle. Let $X_{r,m}$ denote the random variable observed in the mth seasonal period in the rth year, where $m = 1, 2, \ldots, s$. For quarterly data, $s = 4$ and a PAR(1) model may be written in the form

$$X_{r,m} = \phi_{1,m} X_{r,m-1} + Z_{r,m}$$

for $m = 2, 3, 4$, while the value in the first quarter of the year depends on the value in the last quarter of the previous year and so we have

$$X_{r,1} = \phi_{1,1} X_{r-1,4} + Z_{r,1}$$

Procedures are available for choosing the order of the periodic autoregression, for estimating the model parameters and for testing whether there are significant differences between the estimated parameters

---

[6] Logarithms are taken because the seasonality is multiplicative – see Sections 2.3.3 and 2.3.6.

at a particular lag. If not, an ordinary (non-periodic) AR model may be adequate. The author has no practical experience with PAR models and so the reader is referred, for example, to McLeod (1993) and Franses (1996).

3.1.8 Fractional integrated ARMA (ARFIMA) and long-memory models

An interesting variant of ARIMA modelling arises with the use of what is called *fractional differencing*. This idea dates back about 20 years (e.g. Granger and Joyeux, 1980) and is the subject of much current research. In the usual ARIMA(p, d, q) model in (3.1.19), the parameter d is an integer, usually zero or one. *Fractional integrated* ARMA (abbreviated ARFIMA) models extend this class by allowing d to be non-integer. When d is non-integer, then the dth difference $(1 - B)^d X_t$ may be represented by the binomial expansion, namely

$$(1 - B)^d X_t = (1 - dB + d(d - 1)B^2/2! - d(d - 1)(d - 2)B^3/3! \ldots)X_t$$

As such, it is an infinite weighted sum of past values. This contrasts with the case where d is an integer when a finite sum is obtained. Then X_t is called an ARFIMA(p, d, q) model if

$$\phi(B)(1 - B)^d X_t = \theta(B)Z_t$$

where ϕ and θ are polynomials of order p, q, respectively. In other words, the formula is exactly the same as for an ARIMA(p, d, q) model, except that d is no longer an integer. It can be shown (e.g. Brockwell and Davis, 1991, Section 13.2) that the process is stationary provided that $-0.5 < d < 0.5$. If the value of d indicates non-stationarity, then integer differencing can be used to give a stationary ARFIMA process. For example, if an observed series is ARFIMA$(p, d = 1.3, q)$, then the first differences of the series will follow a stationary ARFIMA$(p, d = 0.3, q)$ process.

A drawback to fractional differencing is that it is difficult to give an intuitive interpretation to a non-integer difference. It is also more difficult to make computations based on such a model because the fractional differenced series has to be calculated using the binomial expansion given above. As the latter is an infinite series, it will need to be truncated in practice, say at lag 36 for monthly data. Thus the computation involved is (much) more difficult than that involved in taking integer differences and will result in more observations being 'lost'.[7] Moreover, the parameter d has to be estimated. Although this is not easy, substantial progress has been made, though the literature is technically demanding. Details will not be given here and the reader is referred, for example, to Crato and Ray (1996) and the earlier references therein.

A fractionally integrated variant of SARIMA models can also be defined by allowing both the seasonal and non-seasonal differencing parameters, namely d and D, to take non-integer values (Ray, 1993).

---

[7] For example, only one observation is 'lost' when first differences are taken.

Stationary ARFIMA models, with $0 < d < 0.5$, are one type of a general class of models called *long-memory* models (see Beran, 1992; 1994; and the special issue of *J. Econometrics*, 1996, **73**, No. 1). For most stationary time series models (including stationary ARMA models), the autocorrelation function decreases 'fairly fast', as demonstrated, for example, by the exponential decay in the autocorrelation function of the AR(1) model. However, for some models the correlations decay to zero very slowly, implying that observations far apart are still related to some extent. An intuitive way to describe such behaviour is to say that the process has a *long memory*, or that there is *long-range dependence*. More formally a stationary process with autocorrelation function ρ_k is said to be a *long-memory process* if $\sum_{k=0}^{\infty} |\rho_k|$ does not converge (a more complete technical definition is given by Beran (1994, Definition 2.1)).

Long-memory models have a number of interesting features. Although it can be difficult to get good estimates of some parameters of a long-memory model, notably the mean, it is usually possible to make better forecasts, at least in theory. As regards estimating the mean, the usual formula for the variance of a sample mean is σ^2/N, but this applies to the case of N independent observations having constant variance σ^2. In time-series analysis, successive observations are generally correlated and the variance of a sample mean can be expressed as $\sigma^2 \Sigma_{k=0}^{N-1}(1 - \frac{k}{N})\rho_k/N$. When the correlations are positive, as they usually are, the latter expression can be much larger than σ^2/N, especially for long-memory processes where the correlations die out slowly. In contrast to this result, it is intuitively clear that the larger and longer lasting the autocorrelations, the better will be the forecasts from the model. This can readily be demonstrated, both theoretically and practically (Beran, 1994, Section 8.7), but this topic will not be pursued here.

3.1.9 *Testing for unit roots*

In earlier sections of this chapter, we concentrated on the theoretical properties of the different models, and left most inferential issues until Sections 3.5 and 4.2. However, there is one topic of immediate application to ARIMA-type models, which may conveniently be considered here. This is the problem of testing for unit roots.

A major problem in practice is distinguishing between a process which is stationary and one which is non-stationary. This problem is exacerbated by the possibility that a process may be 'nearly non-stationary' in some general sense. Long-memory stationary processes are arguably in the latter category, as are ordinary AR stationary processes with roots near the unit circle. For example, the (stationary) AR(1) process with parameter 0.95, namely $X_t = 0.95X_{t-1} + Z_t$, will yield data which, for short series, will have properties which look much like those of data generated by a (non-stationary) random walk, namely $X_t = X_{t-1} + Z_t$. The time plots will look

similar, and we see later that the short-term forecasts from the two models
will be close to each other.[8]

Non-stationary, long-memory and other nearly non-stationary processes
will all yield data whose sample autocorrelation function dies out (very)
slowly with the lag and whose sample spectrum will be 'large' at zero
frequency. This makes it hard to distinguish between these very different
types of process on the basis of sample data. In particular, this means that
it will be hard to answer one important question, often asked in regard
to ARIMA models, namely whether the parameter d in (3.1.19) is exactly
equal to one. If it is, then that would mean that a *unit root* is present. Put
another way, if the original data are non-stationary but the first differences
are stationary, then a unit root is said to be present.

There is a large and growing literature in econometrics (e.g. Dejong and
Whiteman, 1993; Stock, 1994; Hamilton, 1994, Chapter 15) on *testing for
a unit root*. There are several types of test designed for different alternative
hypotheses, one example being the so-called *augmented Dickey-Fuller test*,
details of which may be found in Enders (1995, Chapter 4) or Harvey (1993,
Section 5.4). The tests generally take the *null* hypothesis to be that there *is*
a unit root (so that $d = 1$), presumably on the grounds that many economic
series are known to be 'close' to a random walk or ARIMA$(0, 1, 0)$ process.
However, it is not obvious that this is necessarily a sensible choice of null
hypothesis. The statistician, who is interested in fitting an ARIMA model,
will generally make no prior assumption other than that some member of
the ARIMA class is appropriate. Thus, rather than assume (albeit perhaps
supported by a test) that $d = 1$, the statistician is likely to be interested
in assessing an appropriate value for d (and also for p and q, the orders of
the AR and MA parts of the model) without any prior constraints.

The question as to whether it is sensible to test for the presence of
a unit root is a good example of the different ways that statisticians
and econometricians may go about formulating a time-series model – see
Section 3.5. Econometricians tend to carry out a series of tests, not only
for the presence of a unit root, but also for other features such as constant
variance, autocorrelated residuals and so on. In contrast, statisticians are
more likely to choose a model by selecting a general class of models and
then selecting a member of this family so as to minimize a criterion such
as Akaike's Information criterion (AIC). Tests may only be used to check
the residuals from the 'best' model. Bayesian statisticians will also avoid
tests and attempt to assess the strength of evidence as between competing
models by calculating posterior odds ratios (Marriott and Newbold, 1998).

Tests for unit roots have particular problems in that they generally
have poor power, even for moderate size samples, because the alternative
hypothesis is typically 'close' to the null hypothesis and the testing

---

[8] But note that long-term forecasts may be substantially different asymptotically. For
the AR(1) model the long-run forecasts revert to the overall mean, whereas those for
the random walk are all equal to the most recent observed value.

procedures are usually sensitive to the way that lag structure is modelled (Taylor, 1997). It is possible to devise tests with somewhat better power for specific situations (Diebold, 1998, Chapter 10), but some analysts will agree with Newbold et al. (1993), who go so far as to say that "testing for unit autoregressive roots is misguided".

My own view is that choosing an appropriate description of any non-stationary behaviour is crucial in modelling and forecasting, but that a formal test for a unit root can only ever be a small contribution to this task. The fact that we cannot reject a unit root, does not mean that we should necessarily impose one, as, for example, if we want an explicit estimate of the trend rather than difference it away. Conversely, there could still be practical reasons why we might wish to difference our data, even when a unit root is rejected, as, for example, when a model for the differenced data appears to be more robust to unexpected changes. The key question for the forecaster is not whether a unit-root test helps select the 'true model', but whether the chosen model (which we fully expect to be misspecified in some respects) gives better out-of-sample forecasts than alternatives. There is, in fact, some simulation evidence (Diebold and Kilian, 2000) that unit root tests can help to select models that give superior forecasts. Even so, at the time of writing, there are still unresolved questions as to when, or even if, unit root tests can help the forecaster. Thus this complex topic will not be further considered here.

While rejecting the testing for unit roots as the principal tool for assessing non-stationarity, some further remarks on different types of non-stationarity may be helpful. Section 2.3.5 distinguished between various types of trend, notably between a global and a local linear trend. Econometricians make a similar distinction but use a different vocabulary, by referring to what they call *difference-stationary* series, where stationarity can be induced by first differencing, and *trend-stationary* series where the deviations from a deterministic trend are stationary. The random walk is a simple example of a difference-stationary series, as is any ARIMA model with a unit root, meaning that $d \geq 1$. In contrast, if $X_t = a + bt + \varepsilon_t$, where $\{\varepsilon_t\}$ is stationary, then the series is trend-stationary. In the latter case, the trend is deterministic, while econometricians generally say that there is a *stochastic trend* for difference-stationary series. The general consensus is that most economic series are difference-stationary rather than trend-stationary, and there is empirical evidence that difference-stationary models tend to give better out-of-sample forecasts for non-stationary data (Franses and Kleibergen, 1996).

Spurious autocorrelations can readily be induced in a series showing trend, either by mistakenly removing a deterministic trend from difference-stationary data, or by differencing trend-stationary data. This illustrates the importance of identifying the appropriate form of trend so that the appropriate level of differencing may be applied. However, from what is said above, it is clear that it is generally rather difficult to distinguish between the cases (i) $d = 1$, (ii) $0 < d < 1$, (iii) $d = 0$, and (iv) trend-

stationarity. This is perhaps not too surprising given that it is possible
to construct examples where models with different orders of differencing
or with different trend structures can be made in some sense arbitrarily
close. This is illustrated by considering the AR(1) model in (3.1.4) as the
coefficient ϕ tends to one. For $\phi < 1$, we have a stationary model with
$d = 0$, but, when $\phi = 1$, then the model reduces to a difference-stationary
random walk model with $d = 1$. A somewhat more complicated example is
given in Example 3.2.

Example 3.2   This example shows that an ARIMA model with $d = 0$ can
be made 'arbitrarily close' to a model with $d = 1$. Consider the model

$$X_t = \alpha X_{t-1} + Z_t + \beta Z_{t-1}.$$

When $\alpha = 0.95$ and $\beta = -0.9$, the process is a stationary ARMA(1,1)
model so that $d = 0$ when expressed as an ARIMA(1,0,1) model. When
the value of α is changed 'slightly' to $\alpha = 1$, the operator $(1 - B)$ appears
on the left-hand side of the equation so that the process becomes a non-
stationary ARIMA(0,1,1) model with $d = 1$. However, if, in addition,
we now change the value of β 'slightly' to -1, then the character of the
model will change again. The error term is now $(1 - B)Z_t$ which appears
to mean that the process is no longer invertible. However, the operator
$(1 - B)$ appears on both sides of the equation. A general solution is given
by $X_t = Z_t + \mu$, where μ denotes a constant (which disappears on taking
first differences), and this is stationary white noise with a non-zero mean.
With a typical sample size, it would be nearly impossible to distinguish
between the above three cases. For short-term forecasting, it makes little
difference which model is chosen to make forecasts, but, for long-range
forecasts, choosing a stationary, rather than non-stationary, model can be
very influential both on the point forecasts and on the width of prediction
intervals. As the forecast horizon tends to infinity, point forecasts revert
to the overall mean for a stationary series, but not for a non-stationary
one, while the forecast error variance stays finite for a stationary model
but becomes infinite for a non-stationary model.   □

Given the difficulty in distinguishing between different types of stationarity
and non-stationarity, there is much to be said for choosing a forecasting
method which makes few assumptions about the form of the trend, but
is designed to be adaptive in form and to be *robust* to changes in the
underlying model. One class of models which arguably give more robust
forecasts than ARIMA models are state-space models.

3.2 State space models

The phrase 'state space' derives from a class of models developed by control
engineers for systems that vary through time. When a scientist or engineer

tries to measure a signal, it will typically be contaminated by noise so that

$$\text{observation} \;=\; \text{signal} + \text{noise} \qquad (3.2.1)$$

In state-space models the signal at time t is taken to be a linear combination of a set of variables, called *state variables*, which constitute what is called the *state vector* at time t. Denote the number of state variables by m, and the $(m \times 1)$ state vector by $\boldsymbol{\theta}_t$. Then (3.2.1) may be written

$$X_t = \boldsymbol{h}_t^T \boldsymbol{\theta}_t + n_t \qquad (3.2.2)$$

where $\boldsymbol{h}_t$ is assumed to be a known $(m \times 1)$ vector, and n_t denotes the observation error, assumed to have zero mean.

What is the state vector? The set of state variables may be defined as the minimum set of information from present and past data such that the future behaviour of the system is completely determined by the present values of the state variables (and of any future inputs in the multivariate case). Thus the future is independent of past values. This means that the state vector has a property called the *Markov property*, in that the latest value is all that is needed to make predictions.

It may not be possible to observe all (or even any of) the elements of the state vector, $\boldsymbol{\theta}_t$, directly, but it may be reasonable to make assumptions about how the state vector changes through time. A key assumption of linear state-space models is that the state vector evolves according to the equation

$$\boldsymbol{\theta}_t = G_t \boldsymbol{\theta}_{t-1} + \boldsymbol{w}_t \qquad (3.2.3)$$

where the $(m \times m)$ matrix G_t is assumed known and $\boldsymbol{w}_t$ denotes an m-vector of disturbances having zero means. The two equations 3.2.2 and 3.2.3 constitute the general form of a univariate state-space model. The equation modelling the observed variable in (3.2.2) is called the *observation* (or *measurement*) equation, while (3.2.3) is called the *transition* (or *system*) equation.[9] An unknown constant, say δ, can be introduced into a state-space model by defining an artificial state variable, say δ_t, which is updated by $\delta_t = \delta_{t-1}$ subject to $\delta_0 = \delta$. The 'error' terms in the observation and transition equations are generally assumed to be uncorrelated with each other at all time periods and also to be serially uncorrelated through time. It may also be assumed that n_t is $N(0, \sigma_n^2)$ while $\boldsymbol{w}_t$ is multivariate normal with zero mean vector and known variance-covariance matrix W_t. If the latter is the zero matrix, then the model reduces to time-varying regression.

Having expressed a model in state-space form, an updating procedure can readily be invoked every time a new observation becomes available, to compute estimates of the current state vector and produce forecasts. This procedure, called the *Kalman filter*, only requires knowledge of the most recent state vector and the value of the latest observation, and will be described in Section 4.2.3.

There are many interesting special cases of the state-space model. For

---

[9] Actually a set of m equations.

example, the *random walk plus noise* model of Section 2.5.5, also called the *local level* or *steady* model, arises when θ_t is a scalar, μ_t, denoting the current level of the process, while h_t and G_t are constant scalars taking the value one. Then the local level, μ_t, follows a random walk model. This model depends on two parameters which are the two error variances, namely σ_n^2 and $\text{Var}(w_t) = \sigma_w^2$. The properties of the model depend primarily on the ratio of these variances, namely σ_w^2/σ_n^2, which is called the *signal-to-noise ratio*.

In the *linear growth* model, the state vector has two components, $\theta_t^T = (\mu_t, \beta_t)$ say, where μ_t, β_t may be interpreted as the local level and the local growth rate, respectively. By taking $h_t^T = (1,0)$ and $G_t = \begin{bmatrix} 1 & 1 \\ 0 & 1 \end{bmatrix}$, we have a model specified by the three equations

$$X_t = \mu_t + n_t \tag{3.2.4}$$

$$\mu_t = \mu_{t-1} + \beta_{t-1} + w_{1,t} \tag{3.2.5}$$

$$\beta_t = \beta_{t-1} + w_{2,t} \tag{3.2.6}$$

Equation (3.2.4) is the observation equation, while (3.2.5) and (3.2.6) constitute the two transition equations. Of course, if $w_{1,t}$ and $w_{2,t}$ have zero variance, then there is a deterministic linear trend, but there is much more interest nowadays in the case where $w_{1,t}$ and $w_{2,t}$ do *not* have zero variance giving a local linear trend model – see Section 2.3.5.

Harvey's *basic structural model* adds a seasonal index term, i_t say, to the right-hand side of (3.2.4) and adds a third transition equation of the form

$$i_t = -\sum_{j=1}^{s-1} i_{t-j} + w_{3,t} \tag{3.2.7}$$

where there are s periods in one year. The state vector now has $(s+2)$ components, as the transition equations involve the current level, the current trend and the s most recent seasonal indices. None of these state variables can be observed directly, but they can be estimated from the observed values of $\{X_t\}$ assuming the model is appropriate.

The above models have been proposed because they make intuitive sense for describing data showing trend and seasonal variation, and they are in state-space format directly. Many other types of model, including ARIMA models, can be recast into state-space format, and this can have some advantages, especially for estimation. For example, consider the AR(2) model

$$X_t = \phi_1 X_{t-1} + \phi_2 X_{t-2} + Z_t \tag{3.2.8}$$

Given the two-stage lagged dependence of this model, it is not obvious that it can be rewritten in state-space format with (one-stage) Markovian dependency. However, this can indeed be done in several different ways by introducing a two-dimensional state vector which involves the last two observations. One possibility is to take $\theta_t^T = [X_t, X_{t-1}]$ and rewrite (3.2.8) as follows. The observation equation is just

$$X_t = (1, 0)\boldsymbol{\theta}_t \tag{3.2.9}$$

while the transition equation is

$$\boldsymbol{\theta}_t = \begin{bmatrix} \phi_2 & \phi_2 \\ 0 & 1 \end{bmatrix} \boldsymbol{\theta}_{t-1} + \begin{bmatrix} 1 \\ 0 \end{bmatrix} Z_t \tag{3.2.10}$$

This formulation is rather artificial, and we would normally prefer the usual AR(2) formulation, especially for descriptive purposes. However, the one-stage Markovian state-space format does enable the Kalman filter to be applied.

Alternative ways of re-expressing an AR(2) model in state-space form include using the state vector $\boldsymbol{\theta}_t^T = (X_t, \phi_2 X_{t-1})$ or $\boldsymbol{\theta}_t^T = (X_t, \hat{X}_{t-1}(1)$. Note that two of the state vectors suggested above *are* observable directly, unlike the state vectors in earlier trend-and-seasonal models.

The lack of uniqueness in regard to state vectors raises the question as to how the 'best' formulation can be found, and what 'best' means in this context. We would like to find a state vector which summarizes the information in the data set in the best possible way. This means, first of all, choosing an appropriate dimension so as to include all relevant information into the state vector, but avoid including redundant information. Given a set of data, where the underlying model is unknown, there are various ways of trying to find an appropriate state vector (see, for example, Aoki, 1987, Chapter 9) which essentially involve looking at the dimension of the data using techniques like canonical correlation analysis, where we might, for example, look for relationships between the set of variables (X_t, X_{t-1}, X_{t-2}) and $(X_{t+1}, X_{t+2}, X_{t+3})$. Details will not be given here, because my own experience has been with formulating a model, such as Harvey's Basic Structural Model, using common sense and a preliminary examination of the data.

In the general state-space model defined in (3.2.2) and (3.2.3), it may seem unnatural to assume that the deviations in the transition equations (the w_t's) should be independent with respect to the deviation n_t in the observation equation. An alternative possibility explored by Ord et al. (1997) is to consider the linear model

$$X_t = \boldsymbol{h}^T \boldsymbol{\theta}_t + n_t \tag{3.2.11}$$

where

$$\boldsymbol{\theta}_t = G\boldsymbol{\theta}_{t-1} + \boldsymbol{\alpha} n_t \tag{3.2.12}$$

and $\boldsymbol{\alpha}$ is a vector of constant parameters. Thus there is only a single source of 'error' here and there is perfect correlation between the deviations in the two equations. It turns out that the Kalman filter for some examples of this model are closely related to various exponential smoothing updating equations (see Section 4.3). The latter are computed using the one and only observed one-step-ahead forecasting error that arises and smoothing parameters which are akin to the components of $\boldsymbol{\alpha}$.

The different state-space formulations cover a very wide range of models and include the so-called *structural* models of Harvey (1989) as well as the *dynamic linear* models of West and Harrison (1997). The latter enable Bayesians to use a Bayesian formulation. The models called *unobserved component models* by econometricians are also of state-space form.

Except in trivial cases, a state-space model will be non-stationary and hence will not have a time-invariant ac.f. Thus state-space models are handled quite differently from ARIMA models in particular. State-space models deal with non-stationary features like trend by including explicit terms for them in the model. In contrast, the use of ARIMA models for non-stationary data involves differencing the non-stationarity away, so as to model the differenced data by a stationary ARMA process, rather than by modelling the trend explicitly. The advantages and disadvantages of state-space models are considered further in Chapters 4 and 6.

Alternative introductions to state-space models are given by Chatfield (1996a, Chapter 10) and Janacek and Swift (1993). A more advanced treatment is given by Aoki (1987) and Priestley (1981).

3.3 Growth curve models

The general form of a growth curve model is

$$X_t = f(t) + \varepsilon_t \qquad (3.3.1)$$

where $f(t)$ is a deterministic function of time only and $\{\varepsilon_t\}$ denotes a series of random disturbances. The ε_t's are often assumed to be independent with constant variance, but in some situations both assumptions may be unwise as the disturbances may be correlated through time and/or their variance may not be constant but may, for example, depend on the local mean level. The function $f(t)$ could be a polynomial function of time (e.g. $a + bt$), or a Gompertz or logistic curve (see Section 2.3.5). In the latter cases, the model is not only non-linear with respect to time, but also non-linear with respect to the parameters. This makes such models more difficult to fit and more difficult to superimpose a suitable error structure. Depending on the context, it may be advisable to choose a function which incorporates a suitable asymptote, for example, to model market saturation when forecasting sales.

Although growth curve models are sometimes used in practice, particularly for long-term forecasting of non-seasonal data, they raise rather different problems than most other time-series models and are not considered further in this book (except for some brief remarks on the construction of prediction intervals in Section 7.5.6). The reader is referred to the review in Meade (1984).

3.4 Non-linear models

The time-series literature has traditionally concentrated on linear methods and models, partly no doubt for both mathematical and practical convenience. Despite their simplicity, linear methods often work well and may well provide an adequate approximation for the task at hand, even when attention is restricted to univariate methods. Linear methods also provide a useful yardstick as a basis for comparison with the results from more searching alternative analyses. However, there is no reason why real-life generating processes should all be linear, and so the use of non-linear models seems potentially promising.

Many observed time series exhibit features which cannot be explained by a linear model. As one example, the famous time series showing average monthly sunspot numbers exhibits cyclic behaviour with a period of approximately 11 years, but in such a way that the series generally increases at a faster rate than it decreases.[10] Similar asymmetric phenomena may arise with economic series, which tend to behave differently when the economy is moving into recession rather than when coming out of recession. As a completely different type of example, many financial time series show periods of stability, followed by unstable periods with high volatility. An example is shown in Figure 3.1.

The time series in Figure 3.1 is the absolute values of daily returns on the Standard & Poor 500 index[11] on 300 successive trading days. The reader should be able to see a period of instability starting at about trading day 75, after the sudden stock market fall[12] on August 31 1998. The series in Figure 3.1 looks stationary in the mean but is non-stationary in variance. Behaviour like this cannot be explained with a linear model, and so non-linear models are usually needed to describe data where the variance changes through time.

Non-linear models have, as yet, been used rather little for serious forecasting but there is increasing interest in such models and they have exciting potential. Thus a brief introduction is given in this section, with some emphasis on the results of empirical forecasting studies. Alternative introductions are given by Chatfield (1996a, Chapter 11), Granger and Newbold (1986, Chapter 10) and Harvey (1993, Chapter 8). More detailed accounts are given by Priestley (1981, Chapter 11; 1988) and by Tong (1990).

---

[10] These data are plotted in Figure 11.1 of Chatfield (1996a). The vertical axis of that graph is much more compressed than might be expected in order that the observer can 'see' the angle of rise and fall more clearly.

[11] This is a weighted index, say I_t, of the top 500 U.S. stocks. Figure 1.1 showed I_t for 90 trading days and Figure 2.2 showed its correlogram. Figure 3.1 shows absolute daily returns measured by $|\log(I_t/I_{t-1})|$ for 300 trading days. The returns can be multiplied by 100 to give a percentage but this has not been done here. The data, part of a much longer series, may be accessed at http://www.bath.ac.uk/~mascc/

[12] This was partially due to concerns about credit failure following a major default on a bonds issue.

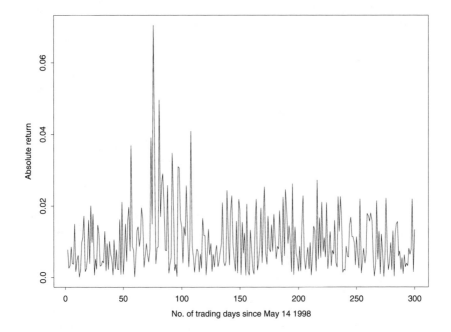

Figure 3.1. *A time plot of the absolute values of daily returns for the S&P 500 index on 300 trading days starting on May 14 1998.*

Questions about non-linearity also arise when considering the possible transformation of an observed variable using a non-linear transformation such as the Box-Cox transformation – see Section 2.3.3. If a linear model can be fitted to the transformed data, then a non-linear model is appropriate for the original data. In particular, a series which shows multiplicative seasonality can be transformed to additive seasonality by taking logs which can then be handled using linear methods. However, the multiplicative model for the original data will be non-linear.

A simple, but important, tool for spotting non-linearities is a careful inspection of the time plot. Features such as asymmetric behaviour and changing variance will be self-evident provided the scales of the time plot are chosen carefully, especially when allied to expert contextual knowledge. However, note that it can be difficult to distinguish between data from (i) a non-linear model, (ii) a linear model with normally distributed disturbances to which some outliers have been added, and (iii) a linear process with disturbances that are not normally distributed.

It is possible to supplement a visual inspection of the time plot with one or more tests for linearity, whose power depends on the particular type of non-linearity envisaged for the alternative hypothesis (e.g. Brock and

Potter, 1993; Patterson and Ashley, 2000). These tests generally involve looking at the properties of moments of $\{X_t\}$ which are higher than second-order, or (working in the frequency domain) by looking at the Fourier transform of suitable higher moments (the so-called cumulants), which are called *polyspectra*. All polyspectra which are higher than second-order vanish when the process is Gaussian. The tests can be made robust to outliers and will then tend to choose linearity more often than would otherwise be the case. The author has little experience with such tests which tend to be used more by econometricians on long financial series.

What is a non-linear model?

Given that we are talking about non-linear models, the reader might reasonably expect that such models can be clearly defined. However, there is no clear consensus as to exactly what is meant by a *linear* stochastic time-series model and hence no consensus as to what is meant by a *non-linear* model. The *general linear process* arises when the value of a time series, say X_t, can be expressed as a (possibly infinite but converging) linear sum of the present and past values of a purely random process so that it is an MA(∞) process – see (3.1.14). This class of models includes stationary AR, MA, and ARMA models. Many state-space models are also generally regarded as linear, provided the disturbances are normally distributed, $\mathbf{h_t}$ is a constant known vector and G_t, W_t are constant known matrices in the notation of Section 3.2. The status of (non-stationary) ARIMA models is not immediately obvious. Because of the non-stationarity, they cannot be expressed as a general linear process, but they look linear in other respects. A further complication is that it is possible to have models which are *locally linear*, but *globally non-linear* (see Section 2.3.5). Thus, rather than try to define linearity and non-linearity precisely, it may be more fruitful to accept that there may be no clearcut distinction between linear and non-linear models, and that in some respects, it is possible to move gradually away from linearity towards non-linearity.

The situation in regard to linearity is much clearer in regard to forecasting *methods*, as opposed to models. A linear forecasting *method* is one where the h-steps-ahead forecast at time t can be expressed as a linear function of the observed values up to, and including, time t. As well as exponential smoothing methods (see Section 4.3), this applies to minimum mean square error (abbreviated MMSE) forecasts derived from stationary ARMA models with known parameters (see Section 4.2.1). Moreover, MMSE forecasts from (non-stationary) ARIMA models (assuming known model parameters) will also be linear functions of past data. This suggests that it might be possible to define a linear model as any model for which MMSE forecasts are linear functions of observed data. However, while this is a necessary condition, it is not sufficient, because some models give linear prediction rules while exhibiting clear non-linear properties in other respects.

To avoid getting into a sterile debate on the above issue, this section restricts attention to certain classes of model that are conventionally regarded as non-linear models, even though some of them have some linear characteristics, while excluding some models that have non-linear characteristics.

What sort of white noise?

As for linear models, some form of white noise is usually an integral building brick of non-linear models, but we will see that it can now be crucial to distinguish between *independent* and *uncorrelated* noise. In Section 2.5.1, *white noise* (or a *purely random process*) was defined to be a sequence of uncorrelated, identically distributed random variables. This is sometimes called *uncorrelated white noise* (UWN) to distinguish it from *strict white noise* (SWN) when successive values are assumed to be independent – a stronger condition. When successive values follow a multivariate normal distribution, the noise is generally described as Gaussian, rather than normal, and in this case zero correlation implies independence so that Gaussian UWN is SWN. However, with non-linear models, distributions are generally non-normal and zero correlation need not imply independence. While UWN has known second-order (linear) properties (constant mean and zero autocorrelations), nothing is specified about the non-linear properties of such series. In particular, although $\{X_t\}$ may be UWN, the series of squared observations $\{X_t^2\}$ need not be. Only if $\{X_t\}$ is SWN will $\{X_t^2\}$ be UWN. Given that non-linear models often involve moments higher than second order, it is necessary to assume that noise is SWN in order to make progress.

When reading the literature about non-linear models, it will also be helpful to know what is meant by a *martingale* and a *martingale difference*. Consider a series of random variables, $\{X_t\}$, and let D_t denote the observed values of X_t available at time t, namely $\{x_t, x_{t-1}, x_{t-2}, \ldots\}$. Then $\{X_t\}$ is called a martingale if $E[X_{t+1} \mid D_t]$ is equal to the observed value of X_t, namely x_t. Thus future expected values do not depend on past values of the series but only on the present value. Following on from this definition, a series $\{Y_t\}$ is called a martingale difference (MD) if $E[Y_{t+1} \mid D_t] = 0$. This last result can readily be explained by letting $\{Y_t\}$ denote the first differences of a martingale, namely $Y_t = X_t - X_{t-1}$, in which case $E[Y_{t+1} \mid D_t] = E[X_{t+1} - X_t \mid D_t] = E[X_{t+1} \mid D_t] - E[X_t \mid D_t] = x_t - x_t = 0$. An MD is like UWN except that it *need not have constant variance*. If an MD is Gaussian and has constant variance, then it is SWN.

This section briefly introduces some classes of non-linear stochastic time-series models of particular interest at the current time. Further details may be found in Tong (1990) and in other references cited below. Throughout this section, we impose the stronger condition that the noise sequence (variously denoted by $\{Z_t\}$ or $\{\varepsilon_t\}$) is independent rather than just uncorrelated, so that it is SWN.

3.4.1 Non-linear autoregressive processes

An obvious way to generalize the (linear) autoregressive model of order p
is to assume that

$$X_t = f(X_{t-1}, X_{t-2}, \ldots, X_{t-p}) + Z_t \qquad (3.4.1)$$

where f is some non-linear function and $\{Z_t\}$ denotes strict white noise.
This is called a *Non-Linear AutoRegressive* model of order p (abbreviated
NLAR(p)). Note that the 'error' term is assumed to be additive. For
simplicity consider the case $p = 1$. Then we can rewrite (3.4.1) as

$$X_t = g(X_{t-1})X_{t-1} + Z_t \qquad (3.4.2)$$

where g is some non-constant function so that the overall function of the
previous value is non-linear. It can be shown that a sufficient condition
for model (3.4.2) to be stable is that g must satisfy the constraint that
$|g(x)| < 1$, at least for large $|x|$. A model such as

$$X_t = \alpha X_{t-1}^2 + Z_t, \qquad (3.4.3)$$

which does not satisfy this condition, will generally be explosive (unless Z_t
is constrained to an appropriate finite interval).

Various models have been proposed which allow the parameter(s) of the
AR model to change through time. For example, we could let

$$X_t = \alpha_t X_{t-1} + Z_t$$

where the autoregressive coefficient α_t itself follows an AR(1) process

$$\alpha_t = \gamma + \beta \alpha_{t-1} + \varepsilon_t$$

where γ, β are constants and $\{\varepsilon_t\}$ denotes a second strict white noise
sequence independent of $\{Z_t\}$. Such models are called *time-varying
parameter models* (e.g. see Nicholls and Pagan 1985). In the case when
$\beta = 0$, the above model reduces to what is sometimes called a *random
coefficient model*.

Another class of models arises if we assume that the function f in (3.4.1)
is piecewise linear and allow the parameters to be determined partly by past
data. This leads to the idea of a *threshold autoregressive model* (abbreviated
TAR model). A simple first-order example of a TAR model (with zero mean
for extra simplicity) is

$$X_t = \begin{cases} \alpha_{(1)}X_{t-1} + Z_t & \text{if } X_{t-1} < r \\ \alpha_{(2)}X_{t-1} + Z_t & \text{if } X_{t-1} \geq r \end{cases} \qquad (3.4.4)$$

where $\alpha_{(1)}, \alpha_{(2)}, r$ are constants and $\{Z_t\}$ denotes strict white noise. This
is like an AR(1) model, but the AR parameter depends on whether X_{t-1}
exceeds the value r called the *threshold*. The AR parameter is $\alpha_{(1)}$ below the
threshold, but $\alpha_{(2)}$ above the threshold. This feature makes the model non-
linear. The model can readily be generalized to higher-order autoregressions
and to more than one threshold. Tong calls a TAR model *self-exciting* (and

uses the abbreviation SETAR) when the choice from the various sets of possible parameter values is determined by just one of the past values, say X_{t-d} where d is the *delay*. In the above first-order example, the choice is determined solely by the value of X_{t-1} and so the model is indeed self-exciting with $d = 1$.

Some theory for threshold models is presented by Tong (1990) and Priestley (1988). In general this theory is (much) more difficult than for linear models. For example, it is quite difficult to give general conditions for stationarity or to compute autocovariance functions. For the first-order model given above, it is intuitively reasonable that the model is likely to be stationary if $|\alpha_{(1)}| < 1$ and $|\alpha_{(2)}| < 1$, and this is indeed a sufficient condition. Proving it is something else.

Another feature of many threshold models is the presence of periodic behaviour called a *limit cycle*. This type of behaviour has links with the behaviour of solutions of some non-linear differential equations in that, if the noise process is 'switched off', then one solution of the process equation has an asymptotic periodic form. This means that, if we plot x_t against x_{t-1}, or more generally against x_{t-k}, then the nature of this cyclic behavior should become obvious. Even when noise is added, the cyclic behaviour may still be visible. The plots of x_t against appropriate lagged values are sometimes called *phase diagrams* or *phase portraits*. They are a useful tool not only for identifying threshold models, but also for investigating other linear and non-linear models. For example, if the plot of x_t against x_{t-1} looks roughly linear, then this indicates an AR(1) model may be appropriate, whereas if the plot looks piecewise linear, but non-linear overall, then a threshold model may be appropriate.

Statistical inference for non-linear models in general, and threshold models in particular, may be substantially different in character from that for linear models. A basic difference is that the (conditional) sum of squared errors surface may no longer be concave. For threshold models, the presence of the threshold parameter creates discontinuities, which can make optimization difficult. It can be helpful to distinguish between what may be called 'regular parameters', such as $\alpha_{(1)}$ and $\alpha_{(2)}$ in the threshold model in (3.4.4), and what may be called 'structural parameters', such as the threshold, r, in (3.4.4), the delay, d, and the orders of the AR models in each of the threshold regions. The latter quantities are all one for the threshold model in (3.4.4). If the structural parameters, especially the value of the threshold, are known, then the sum of squares surface is well behaved with respect to the other (regular) parameters. Then each of the component threshold models can be fitted, using the appropriate subset of observations, to estimate all the regular parameters. However, identifying sensible values for the structural parameters, and then estimating them, is much more tricky, and will not be discussed here. The computation of forecasts for a threshold model is discussed in Section 4.2.4.

Several applications of TAR models have now appeared in the literature. For example, Chappell et al. (1996) analysed exchange rates within the

European Union. These rates are supposed to stay within prescribed bounds so that thresholds can be expected at the upper and lower ends of the allowed range. The threshold model led to improved forecasts as compared with the random walk model. Another interesting example, using economic data, is given by Tiao and Tsay (1994). Although the latter authors found little improvement in forecasts using a threshold model, the modelling process led to greater economic insight, particularly that the economy behaves in a different way when going into, rather than coming out of, recession. Montgomery et al. (1998) compared various non-linear models (including TAR models) with various linear models for forecasting the U.S. unemployment rate. They found little difference in overall out-of-sample forecasting accuracy but the non-linear models were better in periods of sudden economic contraction (but worse otherwise!). Computing forecasts for threshold models more than one step ahead is not easy and Clements and Smith (1997) report a simulation study on a first-order SETAR model which compares different ways of computing the forecasts. Even assuming the analyst knows that the delay, d, is one (normally this would also need to be estimated), the results indicate that forecasts from a linear AR model are often nearly as good, and sometimes better, than those from the SETAR model. Similar findings are presented by Clements and Krolzig (1998) for both real and simulated data. It really does appear that linear models are robust to departures from non-linearity in regard to producing (out-of-sample) forecasts, even when a non-linear model gives a better (in-sample) fit.

TAR models have a discontinuous nature as the threshold is passed, and this has led researchers to consider alternative ways of allowing the AR parameter to change. *Smooth Threshold AutoRegressive* models (abbreviated STAR models) were proposed by Tong (1990) so as to give a smooth continuous transition from one linear AR model to another, rather than a sudden jump. A more recent review of this class of models is given by Teräsvirta (1994), but note that the latter author uses the description *Smooth Transition AutoRegressive* model which leads to the same acronym (STAR).

3.4.2 Some other non-linear models

The *bilinear* class of non-linear models may be regarded as a non-linear extension of the ARMA model, in that they incorporate cross-product terms involving lagged values of the time series and of the disturbance process. A simple example is

$$X_t = \alpha X_{t-1} + \beta Z_{t-1} X_{t-1} + Z_t \qquad (3.4.5)$$

where α and β are constants and $\{Z_t\}$ denotes strict white noise with zero mean and variance σ_Z^2. This model includes one ordinary AR term plus one cross-product term involving Z_{t-1} and X_{t-1}. The second term is the non-linear term.

Some theory for bilinear models is given by Granger and Andersen (1978) and Subba Rao and Gabr (1984). As for threshold models, this theory is generally much more difficult than for linear models, and it is not easy to give general conditions for stationarity or compute autocovariance and autocorrelation functions. It can be shown, for example, that the first-order model given above is stationary provided that $\alpha^2 + \sigma_Z^2 \beta^2 < 1$, but the proof is tricky to say the least, as is the calculation of autocorrelations. Estimation is also tricky, especially when the structural parameters (e.g. the 'orders' of the model) are unknown – see, for example, Priestley (1988, Section 4.1). Computing forecasts is easier than for threshold models – see Section 4.2.4 – but still not as easy as in the linear case.

Bilinear models are like many other non-linear models in that there is no point in looking only at the second-order properties of the observed time series, and hoping that this will identify the underlying model, because it won't! Consider, for example, the bilinear model

$$X_t = \beta Z_{t-1} X_{t-2} + Z_t.$$

After some difficult algebra, it can be shown (Granger and Andersen, 1978) that $\rho(k) = 0$ for all $k \neq 0$. If the analyst simply inspects the sample autocorrelation function in the usual way, then this would suggest to the unwary that the underlying process is uncorrelated white noise (UWN). However, if we examine the series $\{X_t^2\}$, then its autocorrelation function turns out to be of similar form to that of an $ARMA(2,1)$ model. Thus this bilinear model is certainly not strict white noise (SWN), even if it appears to be UWN. Clearly, the search for non-linearity must rely on moments higher than second order. As noted earlier, one general approach is to look at the properties of both $\{X_t\}$ and $\{X_t^2\}$. If both series appear to be UWN, then it is reasonable to treat $\{X_t\}$ as SWN (though this is not a conclusive proof).

Although bilinear models have some theoretical interest, they are perhaps not particularly helpful in providing insight into the underlying generating mechanism of a given time series. Moreover, although they sometimes give a good fit to data (e.g. the sunspots data – see Priestley, 1988, Section 4.1), this does not guarantee good out-of-sample forecasts. De Groot and Würtz (1991) demonstrate that bilinear models are unable to capture the cyclic behaviour of the sunspots data so that out-of-sample forecasts become unstable. Other applications of bilinear models to forecasting have been rather rare.

Another general class of non-linear models are the *state-dependent models* described by Priestley (1988), which can be thought of as locally linear ARMA models. They have also been used rather rarely in forecasting and will not be pursued here.

Finally, we mention *regime-switching models* (e.g. Harvey, 1993, Section 8.6; Hamilton, 1994, Chapter 22), where the generating mechanism is different at different points in time, depending on which regime the process is in. The regime cannot be observed directly (e.g. economic 'expansion' or 'contraction'), although it can be inferred from data. For example, the description *hidden Markov process* is used to describe a process that switches between several different AR processes, with transition probabilities that depend on the previous unobservable regime, or state. When there are two states, the process is called a two-state Markov switching AR model. The model is similar in spirit to a TAR model in that the AR parameters depend on the current (unobservable) regime. When the model changes, it is said to *switch between regimes*. While it is generally easy to compute forecasts within a given regime, there is now the added complication of having to forecast the appropriate regime in some future time period.

If it is possible to compute the probability that a regime-switching process will be in any particular regime in some future period, the analyst may forecast conditional on the regime with the highest probability, or (better) compute unconditional forecasts by taking the weighted average over the conditional forecast for each regime with respect to the probability of being in that regime. Sadly, the examples presented by Clements and Krolzig (1998) and Montgomery et al. (1998) are not encouraging. Dacco and Satchell (1999) are even more discouraging.

3.4.3 Models for changing variance

A completely different class of models are those concerned with modelling *changes in variance*, often called *changes in volatility* in this context. The objective is not to give better point forecasts of the observations in the given series but rather to give better estimates of the (local) variance which in turn allows more reliable prediction intervals to be computed. This can lead to a better assessment of risk.

The estimation of local variance is especially important in financial applications, where observed time series often show clear evidence of changing volatility when the time plot is appropriately presented.[13] An example was shown in Figure 3.1 using the (compounded) returns on the Standard & Poor 500 index. For such series, large absolute values tend to be followed by more large (absolute) values, while small absolute values are often followed by more small values, indicating high or low volatility, respectively.

Many financial series, from which series of returns are calculated, are known to be (very close to) a random walk. For example, the raw data for the Standard & Poor 500 index, plotted in Figure 1.1, were analysed in Example 2.2 and the correlograms suggested a random walk. This means that there is little scope for improving point forecasts of the original series. However, accurate assessment of local variance, by modelling data such

---

[13] The series may also give 'significant' results when a test for homoscedastic variance is carried out (e.g. Gouriéroux, 1997, Section 4.4), although such tests are not really recommended by this author. If the effect is not obvious in the time plot, it is probably not worth modelling.

as that in Figure 3.1, should allow more accurate prediction intervals to be calculated. Note that financial series are often found *not* to be normally distributed (another indication of non-linearity), and this means that prediction intervals may not be symmetric.

Suppose we have a time series from which any trend and seasonal effects have been removed and from which linear (short-term correlation) effects may also have been removed. Thus $\{Y_t\}$ could, for example, be the series of residuals from a regression or autoregressive model for the first differences of a financial time series such as (natural log of) price. The notation $\{Y_t\}$, rather than $\{X_t\}$, is used to emphasize that models for changing variance are rarely applied directly to the observed data. The derived series, $\{Y_t\}$, should be (approximately) uncorrelated but may have a variance which changes through time. Then it may be represented in the form

$$Y_t = \sigma_t \varepsilon_t \qquad (3.4.6)$$

where $\{\varepsilon_t\}$ denotes a sequence of independent random variables with zero mean and unit variance and σ_t may be thought of as the local conditional standard deviation of the process. The ε_t may have a normal distribution but this assumption is not necessary for much of the theory. In any case the unconditional distribution of data generated by a non-linear model will generally be fat-tailed rather than normal. Note that σ_t is not observable directly.

Various assumptions can be made about the way that σ_t changes through time. The *AutoRegressive Conditionally Heteroscedastic model* of order p, abbreviated ARCH(p), assumes that σ_t^2 is linearly dependent on the last p squared values of the time series. Thus Y_t is said to follow an ARCH(1) model if it satisfies (3.4.6) where the conditional variance evolves through time according to the equation

$$\sigma_t^2 = \gamma + \alpha y_{t-1}^2 \qquad (3.4.7)$$

where the constant parameters γ and α are chosen to ensure that σ_t^2 must be non-negative, and y_{t-1} denotes the observed value of the derived series at time $(t-1)$. Note the absence of an 'error' term in (3.4.7). If Y_t can be described by an ARCH model, then it is uncorrelated white noise, but it will not be strict white noise. For example, if $\{Y_t\}$ is ARCH(1), then it can be shown (e.g. Harvey, 1993, Section 8.3) that the autocorrelation function of $\{Y_t^2\}$ has the same form as that of an AR(1) model.

The ARCH model has been generalized to allow linear dependence of the conditional variance, σ_t^2, on past values of σ_t^2 as well as on past (squared) values of the series. The *Generalized* ARCH (or GARCH) model of order (p, q) assumes the conditional variance depends on the squares of the last p values of the series and on the last q values of σ_t^2. For example, the conditional variance of a GARCH(1, 1) model may be written

$$\sigma_t^2 = \gamma + \alpha y_{t-1}^2 + \beta \sigma_{t-1}^2 \qquad (3.4.8)$$

where the parameters γ, α, β must satisfy $(\alpha + \beta) < 1$ for stationarity.

The idea behind a GARCH model is similar to that behind an ARMA model, in the sense that a high-order AR or MA model may often be approximated by a mixed ARMA model, with fewer parameters, using a rational polynomial approximation. Thus a GARCH model can be thought of as an approximation to a high-order ARCH model.

The GARCH$(1,1)$ model has become the 'standard' model for describing changing variance for no obvious reason other than relative simplicity. In practice, if such a model is fitted to data, it is often found that $(\alpha + \beta) \simeq 1$ so that the stationarity condition may not be satisfied. If $\alpha + \beta = 1$, then the process does not have finite variance, although it can be shown that the squared observations are stationary after taking first differences leading to what is called an Integrated GARCH or IGARCH model. Other extensions of the basic GARCH model include Quadratic GARCH (QGARCH), which allows for negative 'shocks' to have more effect on the conditional variance than positive 'shocks', and exponential GARCH (EGARCH) which also allows an asymmetric response by modelling $\log \sigma_t^2$, rather than σ_t^2. These extensions will not be pursued here.

Identifying an appropriate ARCH or GARCH model is difficult, which partially explains why many analysts assume GARCH$(1,1)$ to be the 'standard' model. A (derived) series with GARCH$(1,1)$ variances may look like uncorrelated white noise if second-order properties alone are examined, and so non-linearity has to be assessed by examining the properties of higher order moments (as for other non-linear models). If $\{Y_t\}$ is GARCH$(1,1)$, then it can be shown (e.g. Harvey, 1993, Section 8.3) that $\{Y_t^2\}$ has the same autocorrelation structure as an ARMA$(1,1)$ process. Estimating the parameters of a GARCH model is also not easy. Fortunately, software is increasingly available, either as a specialist package (e.g. S+GARCH) or as an add-on to one of the more general packages (e.g. EViews or RATS).

Forecasting the conditional variance one step ahead follows directly from the model. Forecasting more than one step ahead is carried out by replacing future values of σ_t^2 and of y_t^2 by their estimates. GARCH models have now been used in forecasting a variety of financial variables, where estimation of variance is important in the assessment of risk. These include share prices, financial indices and the price of derivatives such as options to buy a certain share at a pre-specified time in the future. The evidence I have seen indicates that it is often important to allow for changing variance, but that GARCH models do not always outperform alternative models. Sometimes a random walk model for the variance is better than GARCH, while GARCH may not cope well with sudden changes in volatility or with asymmetry. In the latter case, something like EGARCH or a stochastic volatility model (see below) may be better.

The study of ARCH and GARCH models is a fast-growing topic. Alternative introductory accounts are given, for example, by Enders (1995, Chapter 3), Franses (1998, Chapter 7), Harvey (1993, Chapter 8) and Shephard (1996). More advanced details, together with applications in economics and finance, can be found in Bollerslev et al. (1992, 1994), Hamilton (1994, Chapter 21) and Gouriéroux (1997).

An alternative to ARCH or GARCH models is to assume that σ_t^2 follows a stochastic process. This is usually done by modelling $\log(\sigma_t^2)$ to ensure that σ_t^2 remains positive. Models of this type are called *stochastic volatility* or *stochastic variance models*. It seems intuitively more reasonable to assume that σ_t changes stochastically through time rather than deterministically, and the resulting forecasts are often at least as good as those from GARCH models. More details may be found in Harvey (1993, Section 8.4) and Taylor (1994).

Another alternative is to apply intervention analysis (see Section 5.6) to describe sudden changes in variance. For example, Omran and McKenzie (1999) analysed the daily U.K. FTSE All-share index, say I_t, from 1970 to 1997. There is clear evidence that the variance of the returns series, $100 \times \log(I_t/I_{t-1})$, is not constant over time, but that changes in variance can largely be explained by two exceptional shifts in variance during the 1973 oil crisis and the 1987 market crash (the interventions). Outside these periods, the returns can reasonably be described as covariance stationary.

It can be difficult to distinguish between different models for changing variance even with long (greater than 1000) observations. It should be made clear that the point forecasts of time series using models which allow for changing variance will generally be little, if any, better than forecasts from alternative models assuming a constant variance. Rather models which allow for changing variance may be better for predicting *second moments* of the process. Thus models which allow for changing variance need to be checked and compared by forecasting values of Y_t^2 (assuming the conditional mean of Y_t is zero), but this is not easy given that the proportional variability in the squared values is much higher than that of the series itself – see West and Cho (1995). Thus the modelling implications of models for changing variance may be more important than forecasting applications.

Note that McKenzie (1999) has suggested that it is not necessarily appropriate to look at the series of square values, as specified in ARCH-type models, but that some other power transformation may be better. In McKenzie's examples, a power term of 1.25, rather than 2, was indicated.

3.4.4 Neural networks

A completely different type of non-linear model is provided by *Neural networks* (abbreviated NNs), whose structure is thought to mimic the design of the human brain in some sense. NNs have been applied successfully to a wide variety of scientific problems, and increasingly to statistical applications, notably pattern recognition (e.g. Ripley, 1996). The topic is a rapidly expanding research area. NN models are arguably outside the scope of this book, partly because they are not conventional time series models in the sense that there is usually no attempt to model the 'error' component, and partly because they really form a subject in their own right and need much more space to cover adequately. However, in view of the striking, perhaps exaggerated, claims which have been made about their forecasting ability, it seems sensible to say something about them. Alternative introductions, and further references, are given by Faraway and Chatfield (1998), Stern (1996) and Warner and Misra (1996). The introductory chapter in Weigend and Gershenfeld (1994) presents a computer scientist's perspective on the use of NNs in time-series analysis.

A neural net can be thought of as a system connecting a set of inputs to a set of outputs in a possibly non-linear way. In a time-series context, the 'output' could be the value of a time series to be forecasted and the 'inputs' could be lagged values of the series and of other explanatory variables. The connections between inputs and outputs are typically made via one or more hidden layers of *neurons* or *nodes*. The structure of an NN is usually called the *architecture*. Choosing the architecture includes determining the number of layers, the number of neurons in each layer, and how the inputs, hidden layers and output(s) are connected. Figure 3.2 shows a typical NN with three inputs, and one hidden layer of two neurons.

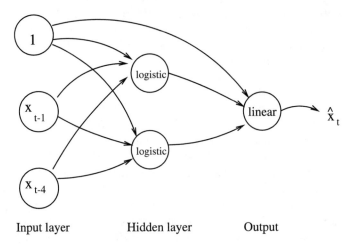

Input layer   Hidden layer   Output

Figure 3.2. *Architecture for a typical NN for time-series forecasting with three inputs (the lagged values at $(t-1)$ and $(t-4)$, and a constant), one hidden layer of two neurons, and one output (the forecast).*

The net in Figure 3.2 is of the usual *feed-forward* type as there are no feedback loops. A suitable architecture for a given problem has to be determined from the context, perhaps using external considerations and perhaps using the properties of the data. Sometimes trial-and-error is needed, for example, to choose a sensible number of hidden neurons. Thus if we want to forecast quarterly data, for example, then it is natural to include the values at lags one and four as inputs when determining the one-step-ahead forecast. In addition, it is usually advisable to include a constant input term which for convenience may be taken as unity. One hidden layer of two neurons, as in Figure 3.2, is usually large enough unless the series is very non-linear. Thus the architecture in Figure 3.2 seems reasonable for forecasting quarterly data, and can be likened to a sort of non-linear (auto)regression model.

In Figure 3.2, each input is connected to both the neurons, and both neurons are connected to the output. There is also a direct connection from the constant input to the output. The 'strength' of each connection is measured by a parameter called a *weight*. There may be a large number of such parameters to estimate. A numerical value is calculated for each neuron at each time period, t, as follows. Let $y_{i,t}$ denote the value of the ith input at time t. In our example, the values of the inputs are $y_{1,t} =$ unity, $y_{2,t} = x_{t-1}$ and $y_{3,t} = x_{t-4}$. Let w_{ij} denotes the weight of the connection between input y_i and the jth neuron. This is assumed to be constant over time. For each neuron, we now calculate a weighted linear sum of the inputs, say $\sum w_{ij} y_{i,t} = v_{j,t}$, for $j = 1, 2$. The analyst then has to choose a function, called an *activation function*, for transforming the values of v_j into a final value for the neuron. This function is typically non-linear. A commonly used function is the *logistic function*, $z = 1/(1 + e^{-v})$, which gives values in the range $(0,1)$. In our example this gives values $z_{1,t}$ and $z_{2,t}$ for the two neurons at each time period, t. A similar operation can then be applied to the values of $z_{1,t}$, $z_{2,t}$ and the constant input in order to get the predicted output. However, the logistic function should not be used at the output stage in time-series forecasting unless the data are suitably scaled to lie in the interval $(0,1)$. Otherwise the forecasts will be of the wrong order of magnitude. Instead, a linear function of the neuron values may be used, which implies the identity activation function at the output stage.

The introduction of a constant input unit, connected to every neuron in the hidden layer and also to the output, avoids the necessity of separately introducing what computer scientists call a *bias*, and what statisticians would call an intercept term, for each relation. Essentially the 'biases' are replaced by weights which measure the strength of each connection from the unit input and so become part of the overall set of weights (the model parameters) which can all be treated in the same way.

For an NN model with one hidden level of H neurons, the general prediction equation for computing a forecast of x_t (the output) using selected past observations, $x_{t-j_1}, \ldots, x_{t-j_k}$, as the inputs, may be written (rather messily) in the form:

$$\hat{x}_t = \phi_o \left(w_{co} + \sum_{h=1}^{H} w_{ho} \, \phi_h \left(w_{ch} + \sum_{i=1}^{k} w_{ih} \, x_{t-j_i} \right) \right) \tag{3.4.9}$$

where $\{w_{ch}\}$ denote the weights for the connections between the constant input and the hidden neurons, for $h = 1, \ldots, H$, and w_{co} denotes the weight of the direct connection between the constant input and the output. The weights $\{w_{ih}\}$ and $\{w_{ho}\}$ denote the weights for the other connections between the inputs and the hidden neurons and between the neurons and the output, respectively. The two functions ϕ_h and ϕ_o denote the the activation functions used at the hidden layer and at the output, respectively.

We use the notation $\mathrm{NN}(j_1, \ldots, j_k; H)$ to denote the NN with inputs at lags $j_1, \ldots, j_k$ and with H neurons in the one hidden layer. Thus Figure 3.2 represents an $\mathrm{NN}(1, 4; 2)$ model.

The weights to be used in the NN model are estimated from the data by minimizing the sum of squares of the within-sample one-step-ahead forecast errors, namely $S = \sum_t (\hat{x}_{t-1}(1) - x_t)^2$, over a suitable portion of the data. This non-linear optimization problem is no easy task. It is sound practice to divide the data into two sections, to fit the NN model to the first section, called the *training set*, but to hold back part of the data, called the *test set*, so as to get an independent check on predictions. Various fitting algorithms have been proposed for NN models, and many specialist packages are now available to implement them. However, even the better procedures may take several thousand iterations to converge, and yet may still converge to a local minimum. This is partly because there are typically a large number of parameters (the weights) to estimate, and partly because of the non-linear nature of the objective function. The NN literature tends to describe the iterative estimation procedure as being a 'training' algorithm which 'learns by trial and error'. Much of the available software used a popular algorithm called *back propagation* for computing the first derivatives of the objective function, so that S may be minimized. The starting values chosen for the weights can be crucial and it is advisable to try several different sets of starting values to see if consistent results are obtained. Other optimization methods are still being investigated and different packages may use different fitting procedures.

The last part of the time series, the *test set*, is kept in reserve so that genuine out-of-sample forecasts can be made and compared with the actual observations.

Equation (3.4.9) effectively produces a *one-step-ahead forecast* of x_t, namely $\hat{x}_{t-1}(1)$, as it uses the actual observed values of all lagged variables as inputs, and they could include the value at lag one. If *multi-step-ahead forecasts* are required, then it is possible to proceed in one of two ways. Firstly, one could construct a new architecture with several outputs, giving forecasts at one, two, three ... steps ahead, where each output (forecast) would have separate weights for each connection to the neurons. Alternatively, the one-step-ahead forecast can be 'fed back' to replace the lag-one value as one of the input variables. The same architecture could then be used to construct the two-step-ahead forecast, and so on. The latter option is usually preferred.

Note that some analysts fit NN models so as to get the best forecasts of the test set data, rather than the best fit to the training data. In this case the test set is no longer 'out-of-sample' in regard to model fitting and so a third section of data should be kept in reserve so that genuine out-of-sample forecasts can be assessed. This does not always happen!

The number of parameters in an NN model is typically much larger than in traditional time-series models, and for a single-layer NN model is given by $p = (k + 2) H + 1$ where k is the number of input variables (excluding the constant) and H is the number of hidden neurons. For example, the architecture in Figure 3.2 (where k and H are both two) contains 9 connections and hence has 9 parameters (weights). The large number of parameters means there is a real danger that model-fitting will 'overtrain' the data and produce a spuriously good fit which does not lead to better forecasts. This motivates the use of model comparison criteria, such as BIC, which penalize the addition of extra parameters. It also motivates the use of an alternative fitting technique called *regularization* (e.g. Bishop, 1995, Section 9.2) wherein the 'error function' is modified to include a penalty term which prefers 'small' parameter values. This is analogous to the use of a 'roughness' penalty term in nonparametric regression with splines. Research is continuing on ways of fitting NN models, both to improve the numerical algorithms used for doing this, and to explore different ways of preventing over-fitting.

A detailed case study of NN modelling is given by Faraway and Chatfield (1998), and a summary of that case study, focussing on forecasting aspects, is given later in Example 8.4. The empirical evidence regarding the forecasting ability of NNs is mixed and is reviewed in Section 6.4.2.

3.4.5 Chaos

The topic of *chaos* is currently a 'hot' research topic in many areas of mathematical science. Some of the results are relevant to non-linear time-series forecasting. Non-linear dynamical systems typically arise in mathematics in the form of discrete-time maps, such as $x_t = f(x_{t-1})$ where f is non-linear, or as differential equations in continuous time. Such equations are usually deterministic (though noise can be added), and so, in principle at least, can be forecast exactly. In practice, this is usually not the case, partly because the system equations may not be known exactly, and partly because the presence of the phenomenon, called *chaos*, means that future behaviour may be critically dependent on small perturbations to initial conditions.

Some deterministic time series, generated in a non-linear way, exhibit behaviour which is clearly deterministic, in that the series tends to a fixed point, called a *limit point*, or to a limit cycle (see Section 3.4.1), regardless of initial conditions. Such series are not chaotic. However, other deterministic non-linear series appear to be 'random' in many respects, and it is series like this that are often called *chaotic* (though this is not meant to be a formal definition of chaos). As we see below, it can be difficult in practice to decide whether an apparently random time series has been generated by a stochastic model or by a chaotic deterministic model or by some combination of the two.

The classic example of a chaotic series is provided by what mathematicians call the *logistic*, or *quadratic, map*. This is defined by

$$x_t = kx_{t-1}(1 - x_{t-1}) \qquad (3.4.10)$$

for $t = 1, 2, 3, \ldots$ with $x_0 \in (0, 1)$. This is a quadratic deterministic equation, and for $0 < k \leq 4$, any resulting series, generated by the equation, will stay within the range $(0, 1)$. For low values of k, the deterministic nature of the series is self-evident after plotting the time plot. For example, if $0 < k < 1$, a series generated by the logistic map will always decline to zero. In the jargon of chaos theory, we say that all possible *trajectories* (meaning all series generated by this model with $0 < k < 1$) are *attracted* to zero – the limit point – and the latter value is called an *attractor*. However, although we have expressed this in the language of chaos theory, such a series is *not* chaotic. However, for values of k approaching 4, the character of the series changes completely and the resulting sequence of observations will look chaotic with no apparent pattern. Indeed, when $k = 4$, it can be shown that a series, generated by the logistic map, has a flat spectrum with the second-order properties of uncorrelated white noise, regardless of the starting value that is used for the series. Thus, although the series is actually deterministic, it will appear to be completely random if examined in the usual way that we try to identify a suitable linear model, namely examining the time plot and the second-order properties such as the correlogram. And yet, if we were to plot x_t against x_{t-1}, the quadratic deterministic nature of the underlying model *will* quickly become apparent, because the plotted points will all lie on a curve. This emphasizes, once again, that methods applicable for linear models do not carry over to the non-linear case.

A chaotic system has the property that a small change in initial conditions will generally magnify through time rather than die out. This is exemplified in the so-called *butterfly effect*, whereby a butterfly flapping its wings may set in train a tropical storm. The sensitivity to initial conditions (the rate at which a small perturbation is magnified) is measured by a quantity called the *Lyapunov exponent*. This will not be defined here, but values greater than zero indicate that a perturbation of the system leads to divergence of the series, so that chaos is present.[14]

The study of chaos leads to much fascinating mathematics and also to many interesting ideas which will not be pursued here. One important concept is that of the *dimension* of a dynamical system. This may be defined in several different ways, none of which are easy to understand. The dimension of a chaotic system is typically non-integer. Another important finding is that chaotic series often tend, not to a simple limit point or cycle, but to an *attractor set* with a complicated geometrical shape. The latter

---

[14] If a system has a stable fixed point, then it is not chaotic and the Lyapunov exponent is less than zero. A series which consists entirely of noise has an infinite exponent.

is called a *strange attractor*. It will typically be composed of a type of set called a *fractal* which has unusual self-similar properties.

Which, if any, of the results on non-linear dynamical systems in general, and chaos in particular, are useful to a statistician? One obvious question is whether it is possible to forecast a chaotic series. Because of sensitivity to initial conditions, long-term forecasting of chaotic series is usually not possible. However, short-term forecasting of low-dimensional chaotic series is often possible, as the uncertainty induced by perturbations is amplified at a finite rate. Thus, the use of a chaotic model equation may give much better forecasts than linear methods in the short-term (e.g. Kantz and Schreiber, 1997, Example 1.1). It obviously helps if the model is known (Berliner, 1991), or there is lots of data from which to formulate the model accurately. Note that some starting values lead to greater instability than others (e.g. Yao and Tong, 1994) so that the width of 'error bounds' on predictions depends on the latest value from which forecasts are to be made.

Unfortunately, the statistician is generally concerned with the case where the model is usually not known a priori, and where the data are affected by noise. This leads to the difficult question as to whether it is possible to identify a model for a deterministic chaotic series, or distinguish between a chaotic series, a stochastic series or some combination of the two. As to the first question, very large samples are needed to identify attractors in high-dimensional chaos. As to the second problem, we must first distinguish two quite distinct ways in which noise can affect a system. For one type of noise, the true state of the system, say x_t, is affected by *measurement error*, which is usually of an additive form, so that we actually observe $y_t = x_t + n_t$, where n_t denotes the measurement error, even though the system equation for the $\{x_t\}$ series remains deterministic. An alternative possibility is that there is noise in the system equation, so that $x_t = f(x_{t-1}, \varepsilon_t)$, where ε_t denotes what is sometimes called *dynamical noise*. Unfortunately, it may not be possible to distinguish between these two types of situation, or between chaotic series and stochastic series, solely on the basis of data. Noise reduction techniques generally aim to separate the observed series into the signal and the remaining random fluctuations, but classical smoothing methods can actually make things worse for chaotic series (e.g. Kantz and Schreiber, 1997, Section 4.5). Thus the forecaster may find it difficult to disentangle the systematic component of a chaotic model from the 'noise' which will almost certainly affect the system.

A few years ago, there were high hopes that the use of chaos theory might lead to improved economic forecasts. Sadly, this has not yet occurred. It seems unlikely that the stock market obeys a simple deterministic model (Granger, 1992), and, while there is strong evidence of nonlinearity, the evidence for chaos is much weaker (Brock and Potter, 1993).[15] Thus, although chaos theory can be applied in the physical sciences and engineering, my current viewpoint is that it is difficult to apply chaos theory

---

[15] This paper includes details of the so-called BDS test for the presence of chaos.

to forecasting stochastic time-series data. However, research is continuing (e.g. Tong, 1995) and it is possible that the position may change in the future.

A non-technical overview of chaos is given by Gleick (1987) while Isham (1993) gives a statistical perspective. There is also some helpful material in Tong (1990, especially Section 2.11). An interesting, readable introduction from the point of view of mathematical physics is given by Kantz and Schreiber (1997).

3.4.6 Summary

In summary, the possible need for non-linear models may be indicated as a result of:

- Looking at the time plot and noting asymmetry, changing variance, etc.
- Plotting x_t against x_{t-1} (or more generally against x_{t-k} for $k = 1, 2, \ldots$), and looking for limit points, limit cycles, etc.
- Looking at the properties of the observed series of squared values, namely $\{x_t^2\}$, as well as at the properties of $\{x_t\}$.
- Applying an appropriate test for non-linearity.
- Taking account of context, background knowledge, known theory, etc.

Non-linear models are mathematically interesting and sometimes work well in practice, notably for long financial time series. However, the fitting procedure is more complicated than for linear models, and it is difficult to give general advice on how to choose an appropriate type of non-linear model. Nevertheless, the search for a suitable model may lead to greater insight into the underlying mechanism even though the accuracy of the resulting forecasts may show little improvement on those from simpler models.

3.5 Time-series model building

As with most statistical activities, time-series analysis and forecasting usually involves finding a suitable model for a given set of data, and a wide variety of univariate models have now been introduced. This final section explains why models are important and how to go about finding a suitable model for an observed time series. As in other areas of statistics, it is relatively easy to look at the theoretical properties of different models, but much harder to decide which model is appropriate for a given set of data.

A model is generally a mathematical representation of reality and can be used for a variety of purposes including the following:

1. It may provide a helpful *description* of the data, both to model the *systematic* variation and the *unexplained* variation (or 'error' component).

2. By describing the systematic variation, a model may help to confirm or refute a theoretical relationship suggested a priori, or give physical insight into the underlying data-generating process. It may also facilitate comparisons between sets of data.

3. The systematic part of the model should facilitate the computation of good point forecasts, while the description of the unexplained variation will help to compute interval forecasts.

This book is primarily concerned with the latter objective, but it is clear that the various objectives are, to some extent, inter-related.

From the outset, it should be realized that a fitted model is (just) an *approximation* to the data. There will be departures from the model to a greater or lesser extent depending on the complexity of the phenomenon being modelled and the complexity and accuracy of the model. A modern economy, for example, is very complex and is likely to require a more complicated model than the sales regime of a single company. In general, the analyst should try to ensure that the approximation is adequate for the task at hand, and that the model contains as few parameters as necessary to do this.

Statistical model building usually has three main stages, namely:

(a) *Model specification* (or *model identification*);

(b) *Model fitting* (or *model estimation*);

(c) *Model checking* (or *model verification* or *model criticism*).

In practice, a model may be modified, improved, extended or simplified as a result of model checking or in response to additional data. Thus model building is generally an iterative, interactive process (see also Section 8.2). Textbooks typically concentrate on stage (b), namely model fitting, both in time-series analysis and in other areas of Statistics. In fact, computer software is now available to fit most classes of statistical model, including a wide variety of time-series models, so that model fitting is generally straightforward. Chapters 3 to 5 give appropriate references on model fitting for particular classes of time-series model and there is no need to replicate such material in this book.

The real problem is *deciding which model to fit in the first place*, and model specification is the difficult stage where more guidance is needed. The literature sometimes gives the impression that a single model is formulated and fitted, but in practice, especially in time-series analysis, it is more usual to formulate a set or class of candidate models and then select one of them. Thus it can be helpful to partition model specification into (i) *model formulation* and (ii) *model selection* and we consider these two aspects of modelling in turn.

3.5.1 Model formulation

Choosing an appropriate model or class of models is as much an art as a science. There is no single approach that is 'best' for all situations, but it is possible to lay down some general guidelines. The analyst will generally have to (i) get as much *background information* as is necessary, (ii) assess *costs*, (iii) *clarify objectives*, and (iv) have a *preliminary look at the data*. This requires a combination of technical and personal skills. The former include knowledge of appropriate statistical theory as well as experience in analysing real time-series data. The latter include communication skills, general knowledge and common sense. Model building will often be a *team effort* which will call on the skills of specialists in the phenomenon being studied as well as statisticians. The single clear message is that the *context* is crucial in determining how to build a model.

Many of the above points are amplified in Granger (1999, Chapter 1), whose main example discusses the difficulties in trying to model the dynamics of deforestation in the Amazon region. Even with a team of experts, who know the context, there are still some variables which are difficult to define or difficult to measure. Any model which results is tentative and approximate and one model is unlikely to be 'best' for all purposes.

In any study, it is important to decide at an early stage which variables should be included. It is unwise to include too many, but no key variables should be omitted. It is also unwise to include variables which are linearly related. For example, it is dangerous to include the sum $(X_1 + X_2)$ when X_1 and X_2 are already included. If the statistician is provided with data without pre-consultation, it is important to find out how accurate they are and, more fundamentally, whether an appropriate set of variables have been recorded.

A key question is the relative importance to be given to theory and to observed data, although in an ideal world there will be no conflict between these two criteria. The analyst should find out whether there is any accepted theory which should be incorporated. For example, known limiting behaviour and any known special cases should agree with any tentative specification. The analyst should also find out what models have been fitted in the past to similar sets of data, and whether the resulting fit was acceptable. Theory based on past empirical work is more compelling than armchair speculation.

Time-series modelling will also generally take account of a *preliminary analysis* of the given set of data, sometimes called *initial data analysis* or *exploratory data analysis* (see Section 2.3). This will suggest what assumptions are reasonable from an empirical point of view, and is, in any case, vital for getting a 'feel' for the data, for cleaning the data and so on. It will also help determine whether there is enough relevant data available to solve the given problem. The *time plot* is particularly important and any modelling exercise must take account of any observed features, such as trend and seasonality. A preliminary examination of the *correlogram* of the data may also be useful at the preliminary stage – see Section 2.6.

One key decision is whether to use a *black-box* or *structural* type of model. The latter will account for specific physical features and requires substantive subject-matter knowledge in order to construct an intelligent model. This makes it difficult to make general remarks on such models. In contrast, black-box models typically require little subject-matter knowledge, but are constructed from data in a fairly mechanical way. An alternative description for them is *empirical* models, and they may give rather little physical insight into the problem at hand. Univariate time-series models are generally regarded by economists as being of black-box character as they make no attempt to explain how the economy works. Neural network models are another class of models which are often implemented in a black-box way (though whether they should be is another matter – see Faraway and Chatfield, 1998).

Of course, different types of model are required in different situations and for different purposes. The best model for (out-of-sample) forecasting may not be the same as the best model for describing past data. Furthermore some models are 'in-between' the two extremes of being fully 'black-box' or completely structural, while other models can be implemented in different ways depending on context. The class of univariate autoregressive (AR) processes is generally regarded as being of a black-box type but still has a useful role in forecasting. In contrast, the multivariate version (called vector autoregressive (VAR) models – see Section 5.3.3) are of a more structural nature and can be fitted in various ways giving more or less emphasis, as appropriate, to prior knowledge.

Whatever decisions are made during modelling, it may be wise to document them with appropriate explanations. Some statistical theory assumes that the model is given a priori, but most modelling is iterative in nature and the class of candidate models may even change completely as more information comes in or existing information is re-evaluated. If the choice of model depends on the same data used to fit the model, then biases may result – see Chapter 8. Analysts often like to think that modelling is *objective* but in reality *subjective* judgement is always needed, and experience and inspiration are important.

3.5.2 Model selection

In some scientific areas, the analyst may formulate a fairly small number of plausible models (e.g. three or four), and then choose between them in some way. However, in time-series modelling it is more usual to proceed by specifying a broad class of candidate models (such as the ARIMA family) and then selecting a model from within that family which is 'best' in some sense. Some analysts prefer to make this selection using *subjective judgment* by examining diagnostic tools such as the correlogram and the sample partial autocorrelation function of the raw data and of suitably differenced series. The correlogram is particularly useful when selecting an appropriate ARMA model, essentially by finding an ARMA model whose theoretical autocorrelation function (ac.f.) has similar properties to the corresponding observed ac.f. – see Section 4.2.2. The *sample partial autocorrelation function* is helpful for indicating the likely order of an autoregressive (AR) model, if such a model is thought appropriate – see Section 3.1.1. Another diagnostic tool, called the *variogram*,[16] has recently been proposed for use in time-series modelling (e.g. Haslett, 1997), but its use is still at an experimental stage.

Rather than rely entirely on applying subjective judgement to diagnostic tools, many analysts like to choose a model by combining subjective skills with a range of helpful, arguably more 'objective', diagnostic procedures. The latter include *hypothesis tests* and the use of *model-selection criteria*. Some comments on the relative utilities of tests and model-selection criteria were made in Section 3.1.9 in the context of testing for a unit root, and doubts were suggested about using hypothesis tests on a routine basis. More generally, many tests have been proposed in the literature, particularly with econometric applications in mind. There are various misspecification tests, such as tests for normality and constant variance, as well as more general tests, such as tests for the presence of seasonality, of non-linearity or of unit roots. It is not uncommon to see a whole battery of tests being applied to the same set of data in the econometric literature but this raises questions about the overall P-value, about how to choose the null hypotheses and about the existence or otherwise of a 'true' model. Some econometricians have suggested that difficulties can be reduced by using a testing procedure with a 'top-down' approach, rather than a 'bottom-up' approach, meaning that we should put more assumptions, based on background theory, into our null model before starting the testing procedure. However, this does not avoid the problems of knowing what null model to choose, of poor power, and of multiple testing. The author confesses to rarely, if ever, using many of the tests now available in the literature and doubts whether it is ever prudent to carry out a series of hypothesis tests in order to make model-selection decisions (see Granger et al., 1995).

The preference of the author (and of many statisticians) is generally to choose a model by using subjective judgement guided by a variety of statistical diagnostic tools. First select a potentially plausible set of candidate models, perhaps based partly on external contextual considerations. Then examine a variety of statistical pointers. For example, the time plot will indicate if trend and seasonal terms are present, in which case they need to be allowed for, either explicitly in the model, or by filtering them away as in ARIMA-differencing. The correlogram and partial ac.f. of various differenced series will help to indicate an appropriate structure if an ARIMA model is contemplated (see Example 3.3). As for the secondary 'error' assumptions, assume normality and constant variance unless the time plot clearly indicates otherwise. I do not usually carry out misspecification tests, given that most procedures are robust to modest

---

[16] This function has been used to analyse *spatial data* for many years. The theoretical variogram exists for both stationary and (some) non-stationary time series, and is therefore more general in some sense than the autocorrelation function.

departures from the usual 'error' assumptions anyway. Instead, if the choice of model is not yet clear (and perhaps even if it appears to be), I may search over the set of candidate models and select the one which is 'best' according to a suitably selected *model-selection criterion*. Several possible criteria are described below. This type of procedure is easy to implement in a fairly automatic way once the candidate models have been selected. A thorough treatment of *model selection* tools is given elsewhere for various classes of time-series model (e.g. de Gooijer et al., 1985; Choi, 1992) and only a brief introduction will be given here. An example illustrating their use in selecting a neural network model is given later in Example 8.4.

What criterion should we use to select a model in the 'best' way? It is not sensible to simply choose a model to give the best fit by minimizing the residual sum of squares or equivalently by maximizing the coefficient of determination, R^2. The latter measures the proportion of the total variation explained by the model and will generally increase as the number of parameters is increased regardless of whether additional complexity is really worthwhile. There is an alternative fit statistic, called adjusted-R^2, which makes some attempt to take account of the number of parameters fitted, but more sophisticated model-selection statistics are generally preferred. Akaike's Information Criterion (AIC) is the most commonly used and is given (approximately) by:

$$\text{AIC} = -2\ln(\text{max. likelihood}) + 2p$$

where p denotes the number of independent parameters estimated in the model. Thus the AIC essentially chooses the model with the best fit, as measured by the likelihood function, subject to a penalty term which increases with the number of parameters fitted in the model. This should prevent overfitting. Ignoring arbitrary constants, the first (likelihood) term is usually approximated by $N\ln(S/N)$, where S denotes the residual (fit) sum of squares, and N denotes the number of observations. It turns out that the AIC is biased for small samples and may suggest a model with a (ridiculously?) high number of parameters as demonstrated empirically later in Example 8.4. Thus a bias-corrected version, denoted by AIC_C, is increasingly used. The latter is given (approximately) by adding the quantity $2(p+1)(p+2)/(N-p-2)$ to the ordinary AIC, and is recommended for example by Brockwell and Davis (1991, Section 9.3) and Burnham and Anderson (1998). An alternative widely used criterion is the Bayesian Information Criterion (BIC) which essentially replaces the term $2p$ in the AIC with the expression $p + p\ln N$. The BIC, like the AIC_C, penalizes the addition of extra parameters more severely than the AIC, and should be preferred to the ordinary AIC in time-series analysis, especially when the number of model parameters is high compared with the number of observations.

When searching for, and selecting a model, it is common to try many different models. This latter activity is sometimes called *data dredging* or *data mining* and is discussed further in Section 8.2. Although statistics, like the AIC and BIC, penalize more complex models, the reader should realize that there is still a danger that trying lots of different models on the same data may give a spuriously complex model that appears to give a good fit, but which nevertheless gives poor out-of-sample predictions. However, despite the dangers inherent in any form of data dredging, I would still recommend choosing a model with the aid of informed judgement, supplemented where necessary with the use of a model-selection criterion, rather than a series of hypothesis tests, for the following reasons:

1. A model-selection criterion gives a numerical-valued ranking of all models, so that the analyst can see if there is a clear winner or, alternatively, if there are several close competing models. This enables the analyst to assess the strength of evidence when comparing any pair of models, and this is more enlightening than the results of testing one particular model, which might simply be recorded as 'significant' or 'not significant'.

2. A model-selection criterion can be used to compare non-nested models, as would arise, for example, when trying to decide whether to compute forecasts using an ARIMA, neural network or econometric model.

3. Hypothesis tests require the specification of appropriate null hypotheses, which effectively means that some models are given more prior weight than others. Tests also assume that a true model exists and that it is contained in the set of candidate models.

3.5.3 Model checking

Having selected a model, it can now be fitted to the data. As noted earlier, this is usually straightforward and nothing further need be said here on *model fitting*, except to note that some classes of model (e.g. neural networks and GARCH models) may require substantial computing.

The next modelling stage is variously called *model checking*, *model verification* or *model criticism* and is arguably as important as model formulation. If any checks fail, the analyst may need to modify the original model.

Model checking involves ensuring that the fitted model is consistent with background knowledge and also with the properties of the given data. As regards the former, the model should be consistent with external knowledge and with known limiting behaviour. As regards empirical properties, a series of checks will be carried out on the data and will typically involve some sort of *residual analysis*. In time-series analysis, the residuals are generally the one-step-ahead forecasting errors, namely

$$e_t = x_t - \hat{x}_{t-1}(1) \qquad (3.5.1)$$

If the model is a good one, then the residuals should form a random series. They can be examined in several ways, both to check the systematic part of the model and also to check the assumptions made about the innovations.

For example, the residuals can be plotted against time over the whole period of fit and treated as a time series in its own right in order to assess the overall fit of the model. The plot may help to indicate, for example, if there is evidence that the model is changing through time. The mean residual provides a check on *forecast bias*; it should be (nearly) zero if forecasts are unbiased. It may also be worth calculating the mean residual over shorter sections of the time series to see, for example, if forecasts are biased at particular times (e.g. during an economic recession). A large individual residual may indicate an outlying observation, which may need to be looked at specially and perhaps adjusted in some way. The autocorrelation function of the residual series provides an overall check on whether a good model has been fitted or whether there is still some structure left to explain. The residual autocorrelations may be examined individually, to see if any exceed $2/\sqrt{N}$ in absolute magnitude, or the analyst can look at the residual autocorrelations as a whole by calculating the sum of squares of the residual autocorrelations up to some suitably chosen lag and seeing if this is larger than expected if the correct model has been fitted. This forms the basis of the Box-Ljung portmanteau lack-of-fit test. Full details of such *diagnostic checks* are given for example by Box et al. (1994, Chapter 8). If the diagnostic checks suggest that the fitted model is inadequate, then the forecasting method based on it will not be optimal.

The above checks are essentially made *in-sample*. In fact, time-series modelling provides an excellent opportunity to look at *out-of-sample* behaviour, especially when prediction is the main objective. In many ways, the production of reliable forecasts provides a more convincing verification of a model than in-sample tests (where there is always the danger that over-fitting may lead to a spuriously good fit).

Out-of-sample predictions can be checked in two ways. One approach is to divide the complete data set into two parts, fit the model to the first part, but keep back the second part, called the *test set*, so that predictions from the model can be compared with the observed values. An alternative approach is by means of *forecast monitoring*, where the (out-of-sample) one-step-ahead forecast errors are looked at one at a time as each new observation becomes available. The same formula is used as for the residuals in (3.5.1), but the forecasts are now made on an out-of-sample basis. These forecast errors are typically plotted one at a time against time and examined on an ongoing basis. If, for example, a method which has been working well suddenly produces a series of one-step-ahead errors which are all positive, then this systematic under-forecasting indicates that the model may have changed. Such a change will usually be evident from the graph, although various graphical devices are available to assist the forecaster. After spotting a change in behaviour, appropriate corrective action can be taken.

The one-step-ahead forecast errors can also be used to provide routine checks on forecasts, with a variety of procedures based on what are often called *tracking signals* (e.g. Gardner, 1983; McLain, 1988). These methods should be used when there are a large number of series to forecast. One possibility is to calculate the cumulative sum, or cusum, of the forecast errors to give what is called the *cusum tracking signal*. When this is plotted against time, an appropriate rule may be used to indicate if the system is 'in control' or if the forecast errors have become biased.

3.5.4 Further comments on modelling

We close with the following general remarks on modelling:

(a) If the time plot shows that a *discontinuity* is present, or that part of the series has different properties to the rest, then it may not be possible to find a single model to satisfactorily describe all, or perhaps even part of, the data. Forecasting then becomes particularly difficult.

(b) If *outliers* are present, it is essential to accommodate them in the model (e.g. with a long-tailed innovation distribution), to adjust them in some way, or to use *robust* estimation and forecasting methods which are not affected too much by departures from the model assumptions.

(c) If the variation is dominated by *trend and seasonality*, then it is often advisable to model these effects explicitly rather than to simply remove the effects by some sort of filtering or differencing, especially if trend and seasonality are of intrinsic interest anyway.

(d) The problems involved in modelling a short series of perhaps 50 observations (e.g. sales figures) are quite different from those involved in modelling long series of several hundred, or even several thousand observations (e.g. daily temperature readings or daily stock prices). With longer series, statistics, such as the correlogram, can be calculated more accurately, non-linear modelling becomes a possibility and it is often easier to assess if the future is likely to be similar to the past. However, if a series is very long, there is a danger that the early part of the data is not relevant to making forecasts from the end of the series, perhaps because of changes in the underlying model. Thus longer series do not necessarily yield better forecasts.

(e) *Univariate or multivariate?* An important question is whether forecasts should be based on a univariate or multivariate model. A multivariate econometric model, for example, is generally much harder to formulate and fit than a univariate model and the required tools are much more complex. In Chapter 5, we will see that there are particular difficulties in the multivariate case when the variables are interdependent rather than exhibiting what may be described as a one-way causal relationship. Nevertheless, the pay-off for successful multivariate modelling can be high.

(f) *Does a true, known model exist?* Traditional statistical inference generally assumes that there is a true model, and that its structure is known. If this is not the case, then theory formally requires that a model be formulated using one set of data, fitted to a second, and checked on a third. Unfortunately, this is generally impossible in time-series analysis where there is usually only a single realization of a given stochastic process. Splitting a single time series into three parts is not the same as collecting three independent samples, and, in any case, a series may be too short to split into three usable sections. Thus the standard time-series model-building paradigm (see Section 8.2) is to formulate, fit *and* check a model on the *same* set of data, and yet most theoretical results ignore the biases which result from this as well as ignoring the more general effects of *model uncertainty* – see Chapter 8. This book adopts the pragmatic approach that all models are approximations, which may, or may not, be adequate for a given problem. Different approximate models may be appropriate for different purposes.

(g) *What is meant by a 'good' model?* A good model should be (i) consistent with prior knowledge; (ii) consistent with the properties of the data; (iii) be unable to predict values which violate known constraints, and (iv) give good forecasts 'out of sample' as well as 'within sample'. A model with many parameters is not necessarily 'good' by these criteria. A more complicated model may give a better within-sample fit but worse out-of-sample forecasts, and may reduce bias but increase variance. Thus a simple model is often to be preferred to a complicated one. Box et al. (1994) use the adjective 'parsimonious' to describe a model containing a relatively small number of parameters, but which still manages to provide an adequate description of the data for the task at hand. The Principle of Parsimony, sometimes referred to as *Occam's Razor*, suggests that models should contain as few parameters as possible consistent with achieving an adequate fit.

(h) *What is the problem?* Traditional statistical inference is primarily concerned with the interesting, but rather narrow, problem of estimating, and/or testing hypotheses about, the parameters of a *pre-specified* family of parameter-indexed probability models. Chatfield (1995b) has argued that statistical inference should be expanded to include the *whole model-building process*. Setting our sights even wider, model building is just part of *statistical problem solving* (e.g. Chatfield, 1995a) where *contextual considerations*, including *objectives*, are critical. While emphasizing the importance of getting good forecasts, we should not forget that there is usually an underlying problem that needs to be solved (e.g. Should we buy a particular currency? How many items of a particular product should we manufacture? What will the temperature be tomorrow?) and we should not lose sight of that problem.

Example 3.3   Modelling the airline data. The airline data are plotted in Figure 2.1 and have been analysed by many authors. The model fitted by Box et al. (1994) to the logarithms of the data is the so-called airline model in (3.1.22), which is a special type of SARIMA model. This model involves taking logs, taking one non-seasonal difference and then one seasonal difference. The use of logs follows from examining the time plot, noting the multiplicative seasonal effect, and trying to transform it to an additive form. The application of one seasonal and one non-seasonal difference follows from examining the correlograms of various differences of the logged series. The correlogram of the raw (logged) data (not shown here) does not come down to zero at low lags and shows little apart from the need for some sort of differencing. The correlogram of the first differences (also not shown) shows a strong seasonal cycle indicating the need for a further seasonal difference. The correlogram of the logged series after taking one seasonal and one non-seasonal difference is plotted in Figure 3.3. 'Spikes' can be seen at lags one month and twelve months only, and these indicate the need for one non-seasonal and one seasonal MA term.

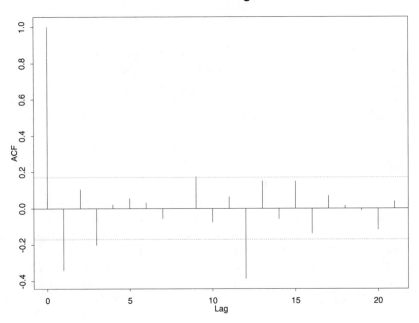

Figure 3.3. *The correlogram of the logged airline data after taking one seasonal and one non-seasonal difference.*

The choice of the airline model is not the only one that might be made, and other analysts might reasonably identify alternative SARIMA models. For example, if you analyse the raw data, rather than the logs, and take one seasonal and one non-seasonal difference, the resulting correlogram looks, if anything, more stationary than that in Figure 3.3. The only 'spike' is at lag one, suggesting an ARIMA$(0, 1, 1) \times (0, 1, 0)_{12}$ or $(1, 1, 0) \times (0, 1, 0)_{12}$

model. However, it turns out that these models give worse values for most model-selection criteria and worse out-of-sample forecasts of the last two years data (if we fit the model to the first 10 years data) than the airline model.

Having selected the airline model, by whatever means, the analyst should then check the fit, by calculating the residuals (the one-step-ahead forecast errors) and then plotting them against time as well as calculating their correlogram. These checks are very easy to carry out with most modern software for fitting ARIMA models. In this case, they indicate the model is adequate.

Of course, the class of SARIMA models is not the only choice that could be made as the set of candidate models. Given the strong trend and seasonal effects, it may seem more natural to fit a model that explicitly models these terms. A traditional trend-and-seasonal model does quite well here, either by fitting multiplicative seasonality to the raw data or additive seasonality to the logarithms. Of course, a model for the raw data is generally more helpful and so this suggests using the former. It turns out that the Holt-Winters method (see Section 4.3.3), which involves finding explicit estimates of trend and seasonality,[17] gives forecasts which are nearly as good as those from the airline model.

Many other models have also been tried on the airline data. State-space models work quite well but neural network models fail to improve on Box-Jenkins and Holt-Winters forecasts (Faraway and Chatfield, 1998).   □

[17] The multiplicative version of Holt-Winters should be used for the raw data in this case and the additive version for the logs.

CHAPTER 4

Univariate Forecasting Methods

"Forecasting is like driving a car blindfolded with help from someone looking out of the rear window"
— Anonymous

This chapter turns attention directly to the topic which is the prime focus of this book, namely forecasting the future values of a time series. In particular, this chapter describes a variety of univariate forecasting methods. Suppose we have observations on a single time series denoted by $x_1, x_2, \ldots, x_N$ and wish to forecast x_{N+h} for $h = 1, 2, \ldots$. A *univariate* forecasting method is a procedure for computing a point forecast, $\hat{x}_N(h)$, based only on past and present values of the given series (possibly augmented with a function of time such as a linear trend).

As discussed more fully in Chapter 6, univariate methods are particularly appropriate when there is a large number of series to forecast, when the analyst's skill is limited or when multivariate methods require forecasts to be made of explanatory variables. We also say more in Chapter 6 about the choice of a particular method and how it will depend on important contextual considerations such as the objectives, the data available (the information set), the quality of data, the software available, and so on. The preliminary questions discussed in Chapter 1 really are vital.

As noted in Section 1.1, it is important to distinguish between a forecasting *method* and a *model*. A model is a mathematical representation of reality, while a method is a rule or formula for computing a forecast. The latter may, or may not, depend on a model. Arising from this distinction, we look in turn at univariate forecasting methods based on fitting a univariate model to the given data (Section 4.2) and then at intuitively reasonable, but essentially ad-hoc, computational procedures (Section 4.3). These two types of method are quite different in character. Before introducing all these forecasting methods, Section 4.1 considers the prediction problem in general and discusses what is meant by a 'good' point forecast.

4.1 The prediction problem

As this chapter deals with univariate forecasting, the following general remarks on the prediction problem are written in the context of univariate forecasts, which has the added advantage of simplicity. Some of the discussion, such as the comments on a choice of loss function, can readily be extended to other types of forecasting, such as the multivariate case.

Given observations on a single time series up to time N, we can denote any univariate forecast of X_{N+h} by $\hat{x}_N(h)$. This could involve a function of time, such as a deterministic linear trend, but we concentrate attention on the more usual case where the forecast can be expressed as some function of the observed data $x_N, x_{N-1}, x_{N-2}, \ldots$, say $\hat{x}_N(h) = g(x_N, x_{N-1}, \ldots)$. Several questions follow immediately such as: (i) What function, g, should we choose? (ii) What properties will it have? (iii) Is it the 'best' forecast? and finally (iv) What is meant by 'best'?

Before answering these questions, the analyst must decide how forecasts should be evaluated and, in particular, what choice of *loss function* is appropriate. A loss function is defined in the following way. Let e denote a forecast error which we may write in words as

$$e = (\text{observed value} - \text{forecast}). \qquad (4.1.1)$$

Then the loss function, say $L(e)$, specifies the 'loss' associated with a forecast error of size e. This function typically has the properties that (i) $L(0) = 0$; and (ii) $L(e)$ is a continuous function which increases with the absolute value of e. It may be symmetric, so that $L(e) = L(-e)$, but could be asymmetric. Two common symmetric loss functions are quadratic loss, where $L(e) = k_1 e^2$, and the absolute error loss function where $L(e) = k_2|e|$, where k_1, k_2 denote constants. An asymmetric loss function arises in production planning if, for example, the loss in losing orders because of understocking exceeds the loss due to overstocking by the same margin. Indeed, the costs attached to over- and under-forecasting are sometimes quite different. However, the loss function is defined, a 'good' forecast may be defined as a forecast which minimizes average loss when averaged over the probability distribution of forecast errors. In practice, it may be difficult to write down a context-specific loss function and so quadratic loss is used more than any other function (perhaps without enough thought being given to this potentially important decision!).

Rather than specify a loss function, some analysts prefer to specify a measure of forecast accuracy in order to evaluate forecasts. The simplest and most widely used measure of forecast accuracy is the *Mean Square Error* (abbreviated MSE), namely $E[(X_{N+h} - \hat{x}_N(h))^2]$. In fact, it can be shown that this measure of accuracy implies a quadratic loss function for departures from prediction and so implicitly assumes a quadratic loss function. In this case, it can readily be shown that the 'best' forecast, in the sense of minimizing MSE, is the conditional expectation of X_{N+h} given the available data at time N, namely[1]

$$\hat{x}_N(h) = E[X_{N+h}|x_N, x_{N-1}, \ldots] \qquad (4.1.2)$$

---

[1] Note that we generally denote a forecast by $\hat{x}_N(h)$ regardless of the criterion and method used to find the forecast. Normally, it will be the conditional expectation given a particular set of data and model, but, if more than one forecast is computed in different ways, then some additional notation will be needed to distinguish between them.

The reader will hopefully find this result intuitively obvious or self-evident, given various well-known results on properties of the mean. For example, in a much simpler context, it is known that the mean of a probability distribution is the value such that the mean square deviation from it is minimized. It is straightforward to extend this result to show that, for time-series data, the conditional expectation of a future value provides minimum MSE forecast accuracy, but the proof is unenlightening and algebraically tedious and will not be repeated here – see, for example, Hamilton (1994, Section 4.1) or Priestley (1981, p. 728). In future we generally assume that the 'best' forecast is the minimum MSE forecast, though it is perfectly reasonable on occasion to use an alternative loss function[2] and hence an alternative measure of forecast accuracy – see Section 6.3.

In order to evaluate a general expression for the conditional expectation of a future value, namely $E[X_{N+h}|X_N, X_{N-1}, \ldots]$, we really need to know the complete conditional distribution of X_{N+h} given $X_N, X_{N-1}, \ldots$ and this is the aim of so-called *density forecasting* – see Section 7.1.4. Alternatively we need to be able to find the joint probability distribution of $\{X_{N+h}, X_N, X_{N-1}, \ldots\}$. This is usually only possible for certain models, notably linear models with normal errors. Partly for this reason, much of the general theory of prediction restricts attention to *linear* predictors of the form

$$\hat{x}_N(h) = \sum_{i=0}^{N-1} w_i x_{N-i} \qquad (4.1.3)$$

where the problem reduces to finding suitable weights $\{w_i\}$ so as to minimize the forecast MSE. Then the forecasts are sometimes called *linear least squares* forecasts. If the process is jointly Gaussian, then it can indeed be shown that a linear predictor is appropriate as the minimum MSE forecast. Even if the process is not Gaussian, a linear predictor may still provide a good approximation to the best MSE forecast.

A variety of results were derived by A. Kolmogorov, N. Wiener, A.M. Yaglom, P. Whittle (Whittle, 1983) and others starting in the 1940's. The material makes challenging reading for the more theoretically inclined and the reader is advised to read the summary in Priestley (1981, Section 10.1). If we know the true correlation function of the process, or equivalently its spectrum, then it is possible to write down the linear least squares predictor, meaning the predictor of linear form which minimizes the expected mean square difference between the actual future value and its forecast. However, this work not only assumes exact knowledge of the correlation function or the spectrum, but may also assume an infinite amount of past data. Neither condition will be satisfied in practice and it is usually only possible to make progress if the estimated spectrum can be readily factorized as a rational function. This happens when the underlying process is an ARMA process

---

[2] Indeed, different users may sometimes quite reasonably choose different loss functions for the same problem depending on their subjective judgement, and this may suggest different measures of forecast accuracy or even lead to different models.

and then it is generally easier to make progress via the ARMA model rather than via the sort of results available in the Wiener-Kolmogorov linear prediction theory. The Wiener-Kolmogorov approach can be thought of as non-parametric in that it is generally based on knowledge of the correlation function or spectrum rather than directly on a fitted model, and so does not lead to a model with the descriptive power to illuminate the situation.

This chapter, like much of the forecasting literature, concentrates on finding 'best' point forecasts using MSE. In practice, we will often want to produce *interval forecasts*, rather than (just) point forecasts, so as to better assess future uncertainty, and this important topic is discussed in Chapter 7. However an ability to calculate good point forecasts is a prerequisite. As regards the use of MSE, the assumption of a quadratic loss function is not always the most sensible choice, and alternative loss functions may lead to alternative forecasting methods.

This chapter also restricts attention to single-period forecasts. Occasionally it may be desirable to forecast the sum of values over a sequence of periods, as for example if we want to plan production over an extended period. These *cumulative forecasts* are typically found simply by adding the relevant single-period point forecasts. However, note that computing the standard error of the uncertainty involved in such a forecast is difficult because the errors in forecasts made from the same time origin will inevitably be correlated.

No numerical illustrations are given in this chapter, as they would not be particularly enlightening. What is important is that the reader should try out selected methods on appropriate data using some of the excellent software now widely available. Having produced some point forecasts, it is always a good idea to plot them as an add-on to the time plot of the data to make sure that they look intuitively reasonable. If they do not, then there is more work to do.

4.2 Model-based forecasting

Suppose we identify a particular model for a given time series, estimate the model parameters and then wish to compute forecasts from the fitted model. We denote the true model by M and the fitted model by M_f. From Section 4.1, the use of a quadratic loss function implies that the best way to compute a forecast is to choose $\hat{x}_N(h)$ to be the expected value of X_{N+h} conditional on the model, M, and on the information available at time N, which will be denoted by I_N. Thus we take

$$\hat{x}_N(h) \quad = \quad E(X_{N+h}|M, I_N) \qquad (4.2.1)$$

For a univariate procedure, I_N consists of $x_N, x_{N-1}, \ldots$ plus the current value of time, namely N. Of course, in practice, we have to use the fitted model, M_f, rather than the true model.

4.2.1 Forecasting with general linear processes

In Section 3.1, we saw that a stationary ARMA process may generally be rewritten as an MA process of possibly infinite order (the Wold representation) in the form

$$X_t = \sum_{j=0}^{\infty} \psi_j Z_{t-j} \qquad (4.2.2)$$

Conditions for convergence are given in Section 3.1, and, when they apply, a sequence of random variables satisfying this equation is called a general linear process. The MMSE forecast of X_{N+h} for the general linear process is given by

$$\hat{x}_N(h) = \sum_{j=h}^{\infty} \psi_j z_{N+h-j} \qquad (4.2.3)$$

(see Box et al., 1994, Equation 5.1.13) where it is assumed, not only that the values of $\{\psi_j\}$ are known, but also that data for the infinite past up to time N, namely $\{x_t,\ t \leq N\}$, are known and hence that the realized values of $\{Z_t,\ t \leq N\}$, denoted by $\{z_t\}$, can be found. Equation (4.2.3) is then intuitively obvious since it follows from (4.2.2) by replacing future values of Z_t by zero, and present and past values of Z_t by their observed values. This merely requires that the MMSE forecast of all future Z's is zero as is obviously the case when the Z's are uncorrelated with mean zero.

As an example of the above result, consider the MA(2) process

$$X_t = Z_t + \theta_1 Z_{t-1} + \theta_2 Z_{t-2}$$

This process has MMSE forecast at time N given by

$$\hat{x}_N(h) = \begin{cases} \theta_1 z_N + \theta_2 z_{N-1} & h = 1 \\ \theta_2 z_N & h = 2 \\ 0 & h \geq 3 \end{cases}$$

More generally, forecasts from an ARMA model are not obtained via (4.2.3), but can be computed directly from the ARMA model equation (which Box et al. (1994) call the *difference equation form*) by replacing (1) future values of Z_t by zero, (2) future values of X_t by their conditional expectation, (3) present and past values of X_t and Z_t by their observed values. For example, the ARMA(1, 1) process

$$X_t = \phi_1 X_{t-1} + Z_t + \theta_1 Z_{t-1}$$

has MMSE forecast at time N given by

$$\hat{x}_N(h) = \begin{cases} \phi_1 x_N + \theta_1 z_N & h = 1 \\ \phi_1 \hat{x}_N(h-1) & h \geq 2 \end{cases} \qquad (4.2.4)$$

Thus the forecasts are generally calculated recursively.

In practice the model parameters are not known exactly but have to be estimated from the data. Moreover, we do not have an infinite number of past observations available as assumed in (4.2.3), while the values of Z_t for $t = 1, \ldots, N$ will not be known exactly, but rather are estimated by the one-step-ahead forecasting errors (the residuals). It seems natural to denote the latter here by $\{\hat{z}_t\}$, though they are denoted by e_t more generally - see for example Equation (3.5.1). These practical points are often given little attention in textbooks, but it must be realized that the prediction formula actually used is not as given in (4.2.3) or in other similar formulae. Instead parameter *estimates* are used together with the observed *residuals*, rather than the true innovations. This is the difference between using the true model, M, and the fitted model, M_f, in (4.2.1). For example, the working version of the ARMA(1, 1) MMSE forecast in (4.2.4) is

$$\hat{x}_N(h) = \begin{cases} \hat{\phi}_1 x_N + \hat{\theta}_1 \hat{z}_N & h = 1 \\ \hat{\phi}_1 \hat{x}_N(h-1) & h \geq 2 \end{cases} \qquad (4.2.5)$$

The use of parameter estimates, and of residuals, increases the prediction error variance, as we will see in Chapter 7.

The above rules for computing MMSE forecasts also apply to (non-stationary) ARIMA models and seasonal ARIMA (or SARIMA) models. For example, the ARIMA(1, 1, 1) model

$$(X_t - X_{t-1}) = \phi_1(X_{t-1} - X_{t-2}) + Z_t + \theta_1 Z_{t-1}$$

may be rewritten as

$$X_t = (1 + \phi_1)X_{t-1} - \phi_1 X_{t-2} + Z_t + \theta_1 Z_{t-1}$$

from which we see that the MMSE forecasts at time N may be calculated recursively by

$$\hat{x}_N(h) = \begin{cases} (1 + \phi_1)x_N - \phi_1 x_{N-1} + \theta_1 z_N & h = 1 \\ (1 + \phi_1)\hat{x}_N(1) - \phi_1 x_N & h = 2 \\ (1 + \phi_1)\hat{x}_N(h-1) - \phi_1 \hat{x}_N(h-2) & h \geq 3 \end{cases}$$

Here again the working version of this formula would actually use parameter estimates, $\hat{\phi}_1$ and $\hat{\theta}_1$, and the observed residual, $\hat{z}_N$.

The forecast resulting from the ARIMA(0, 1, 1) model, $(X_t - X_{t-1}) = Z_t + \theta Z_{t-1}$, is of particular interest. Here we find

$$\begin{aligned} \hat{x}_N(1) &= x_N + \theta z_N \\ &= x_N + \theta[x_N - \hat{x}_{N-1}(1)] \\ &= (1 + \theta)x_N - \theta \hat{x}_{N-1}(1) \\ &= \alpha x_N + (1 - \alpha)\hat{x}_{N-1}(1) \end{aligned} \qquad (4.2.6)$$

where $\alpha = 1 + \theta$. This simple updating formula, which only utilizes the latest observation and the previous forecast, is the basis of many forecasting methods and is called *simple exponential smoothing* (see Section 4.3.1).

General formulae for the forecast error variance for MSE forecasts from a general linear process can be found in terms of the ψ-coefficients when the model is expressed in the MA(∞) form of (4.2.2) – see Section 7.5.2.

4.2.2 The Box-Jenkins forecasting procedure

Section 4.2.1 has shown that it is straightforward to find MMSE forecasts for general linear processes, including ARMA models. The forecasting procedure based on the more general class of ARIMA models is often called the *Box-Jenkins forecasting procedure*, and involves much more than just writing down formulae for point forecasts based on all the different models. Good forecasting depends on finding a suitable model for a given time-series, and the Box-Jenkins approach involves an iterative procedure of (i) *formulating* a plausible model, (ii) *fitting* the model, and (iii) *checking*, and if necessary adjusting, the model. By including ARIMA and SARIMA models, the approach can handle non-stationarity and seasonality. Full details are given in Box et al. (1994) and in many other books, and so the details need not be repeated here.

In brief, the method usually involves (i) looking at the time plot and assessing whether trend and seasonality are present, (ii) taking non-seasonal and seasonal differences until the differenced series is judged to be stationary, (iii) looking at the correlogram and the sample partial ac.f. of the differenced series, and perhaps also at model-selection criteria, in order to identify an appropriate ARIMA model, (iv) estimating the parameters of this model, and (v) carrying out various diagnostic checks on the residuals from the fitted model. If necessary the identified model will be adjusted or alternative models entertained until a model is found which does seem to describe the data adequately. Only then will forecasts be computed as described in Section 4.2.1.

The following points are worth noting:

1. Interpreting the correlogram and the sample partial ac.f. is difficult and the analyst needs to build up considerable experience in choosing a model from these sample functions. Analysts sometimes prefer to circumvent this tricky process by trying a range of models and simply choosing the one which optimizes one of the model-selection statistics defined in Section 3.5.

2. There are several different estimation procedures for fitting ARIMA models which depend primarily on how the first few observations are treated. They include exact maximum likelihood (ML) and conditional least squares (CLS). For example, when CLS is applied to data from an AR(p) process, the first p values of the series are assumed to be fixed. In contrast, exact ML essentially averages over the first few observations in an appropriate way. For long series (several hundred observations), the choice of estimation algorithm usually makes little difference, but for short series, the choice of estimation procedure becomes more important. While asymptotically unbiased, parameter estimates are likely to be biased for short series (e.g. Ansley and Newbold, 1980), and different computer packages, using different estimation routines, can produce parameter estimates with non-trivial differences (Newbold et al., 1994), especially when the model has roots near the unit circle. The analyst

is advised to use software which specifies exactly what estimation procedure is adopted. For short series, exact ML is recommended, while the use of Yule-Walker estimates for AR processes (essentially method of moments estimates) is not recommended, especially when the roots are close to the unit circle.

3. Diagnostic tests, such as the *portmanteau test* based on the ac.f. of the fitted residuals, typically have poor power properties so that models are rarely rejected. This is partly because series tend to be short, but also because time-series models are typically formulated, fitted and checked on the *same* data set – see Chapter 8. It is hardly surprising that the best-fitting model for a given time series is found to be consistent with the same data used to formulate and fit the model.

There are a number of forecasting procedures which are related to, or are arguably a subset of, the Box-Jenkins method. They include *subset autoregression* and Parzen's ARARMA approach. As they are implemented in an automatic way, the latter methods are covered in Section 4.3.4. It can also be argued that some *exponential smoothing* methods are special cases of Box-Jenkins, but this is not a view I subscribe to (see Section 4.4) and they will be considered separately in Section 4.3.

For non-stationary data, the way that trend is removed before applying the Box-Jenkins approach can be vital. If differencing is applied, the *order* of differencing can be crucial. Makridakis and Hibon (1997) show that alternative methods of removing trend, prior to fitting an ARMA model, can lead to better forecasts.

4.2.3 Forecasting with state-space models – the Kalman filter

Point forecasts may readily be obtained for the general state-space model of Section 3.2 by appropriate computations, based on a procedure called the Kalman filter, which is described below. If we knew the exact form of the state-space model, including the exact value of the current state vector, then (3.2.2) suggests we take

$$\hat{x}_N(h) = h_{N+h}^T \boldsymbol{\theta}_{N+h} \qquad (4.2.7)$$

where h_{N+h} is assumed known, and $\boldsymbol{\theta}_{N+h} = G_{N+h} G_{N+h-1} \ldots G_{N+1} \boldsymbol{\theta}_N$, assuming future values of G are also known. In practice, the exact value of $\boldsymbol{\theta}_N$ will not be known and has to be estimated from data up to time N. Then (4.2.7) is replaced by $\hat{x}_N(h) = h_{N+h}^T \hat{\boldsymbol{\theta}}_{N+h}$. In the important special case where h_t and G_t are (known) constant functions, say h and G, then the forecast formula becomes

$$\hat{x}_N(h) = h^T G^h \hat{\boldsymbol{\theta}}_N \qquad (4.2.8)$$

Assuming the structure of the model is known, the computation of forecasts therefore hinges on being able to get good estimates of the current state vector, $\boldsymbol{\theta}_N$. A key property of state-space models is that the latter quantity can readily be obtained with the updating formulae of the *Kalman filter* as each new observation becomes available. The Kalman filter will also give estimates of the variance-covariance matrix of $\hat{\boldsymbol{\theta}}$ to indicate the likely size of estimation errors.[3]

The Kalman filter is usually carried out in two stages. Suppose we have an estimate of $\boldsymbol{\theta}_{t-1}$, say $\hat{\boldsymbol{\theta}}_{t-1}$, based on data up to time $(t-1)$ together with an estimate of its variance-covariance matrix which we denote by $\boldsymbol{P}_{t-1}$. The first stage, called the *prediction stage*, is concerned with forecasting $\boldsymbol{\theta}_t$ using the data up to time $(t-1)$. If we denote the resulting estimate by $\hat{\boldsymbol{\theta}}_{t|t-1}$, then (3.2.3) suggests we use the formula

$$\hat{\boldsymbol{\theta}}_{t|t-1} = G_t \hat{\boldsymbol{\theta}}_{t-1} \qquad (4.2.9)$$

It can be shown that the variance-covariance matrix of this estimate is given by

$$\boldsymbol{P}_{t|t-1} = G_t \boldsymbol{P}_{t-1} G_t^T + W_t \qquad (4.2.10)$$

using the notation of Section 3.2.

When the new observation at time t, x_t, becomes available, the second stage of the Kalman filter, called the *updating stage*, is carried out using the formulae

$$\hat{\boldsymbol{\theta}}_t = \hat{\boldsymbol{\theta}}_{t|t-1} + K_t e_t \qquad (4.2.11)$$

and

$$\boldsymbol{P}_t = \boldsymbol{P}_{t|t-1} - K_t \boldsymbol{h}_t^T \boldsymbol{P}_{t|t-1} \qquad (4.2.12)$$

where

$$e_t = x_t - \boldsymbol{h}_t^T \hat{\boldsymbol{\theta}}_{t|t-1}$$

is the prediction error at time t, and

$$K_t = \boldsymbol{P}_{t|t-1} \boldsymbol{h}_t / [\boldsymbol{h}_t^T \boldsymbol{P}_{t|t-1} \boldsymbol{h}_t + \sigma_n^2]$$

is called the Kalman gain matrix. Further details will not be given here – see, for example, Chatfield (1996, Chapter 10), Harvey (1993, Chapter 4) or Janacek and Swift (1993).

As the name suggests, the Kalman filter is primarily intended for *filtering*, which usually implies getting an estimate of quantities relating to the most recent observation. However it can also be used in prediction, as noted above, and in *smoothing*, which usually implies estimating quantities relating to observations within the body of the data, as for example to interpolate missing observations. The Kalman filter is also now used in a variety of additional statistical applications, such as in updating estimates of the parameters of an ARMA model. All in all, Kalman filtering has become an important statistical tool, with applications outside its original intended use in handling state-space models.

The forecasting procedure, called *Bayesian forecasting* (Pole et al., 1994;

---

[3] The diagonal elements of the variance-covariance matrix are the variances of the different elements of $\hat{\boldsymbol{\theta}}$, while the off-diagonal elements are the covariances between the different elements.

West and Harrison, 1997), is based on a general model, called the *dynamic linear model*. This looks like a state-space model, but forecasts are based on a Bayesian rationale which involves updating prior estimates of model parameters to get posterior estimates. The set of recurrence relationships that result appears to be equivalent to the Kalman filter. The approach is particularly useful for short series when there really is some prior information, and some convincing illustrative examples are given by Pole et al. (1994).

4.2.4 Forecasting with non-linear models

The topic of forecasting with non-linear models has been rather neglected in the literature, partly, no doubt, because it is generally (much) more difficult to compute forecasts from a non-linear model than from a linear model (Lin and Granger, 1994; Tong, 1990, especially Chapter 6). In particular, closed-form analytic formulae only exist in certain special cases for forecasts more than one step ahead. To illustrate the problems, we consider some simple threshold and bilinear models. As in the linear case, the minimum mean square error predictor of X_{N+h} is given by the conditional expectation in (4.1.2), namely

$$\hat{x}_N(h) = E[X_{N+h}|\text{data to time } N]$$

Consider the simple first-order SETAR model in (3.4.4), namely:

$$X_t = \begin{cases} \alpha_{(1)}X_{t-1} + Z_t & \text{if } X_{t-1} < r \\ \alpha_{(2)}X_{t-1} + Z_t & \text{if } X_{t-1} \geq r \end{cases}$$

where $\alpha^{(1)}$, $\alpha^{(2)}$ are constants and $\{Z_t\}$ denotes strict white noise. Suppose we have data up to time N and that x_N happens to be larger than the threshold, r. Then it is obvious from the model that $\hat{x}_N(1) = E[X_{N+1}|\text{data to time } N] = \alpha_{(2)}x_N$, and so the one-step-ahead forecast is easy to find.

However, finding the two-steps-ahead forecast for the above model is much more complicated, as it will depend on whether the next observation happens to exceed the threshold. The expectation that arises is algebraically intractable, as we would need to take expectations over future error terms and the corresponding thresholds. In practice, analysts sometimes use an ad hoc 'deterministic' rule, whereby future errors are set equal to zero. Thus, for the above model, if we find, for example, that $\hat{x}_N(1)$ happens to be smaller than the threshold, r, then we could take $\hat{x}_N(2) = \alpha_{(1)}\hat{x}_N(1) = \alpha_{(1)}\alpha_{(2)}x_N$. However, this is only an approximation to the true conditional expectation, namely $E[X_{N+2}|\text{data to time } N]$. More generally, when the delay is d, it is easy to write down exact formulae for the conditional expectation up to d steps ahead, but forecasts more than d steps ahead cannot generally be found analytically.

Consider now the simple first-order bilinear model from Section 3.4.2, namely:

$$X_t = \alpha X_{t-1} + \beta Z_{t-1}X_{t-1} + Z_t \qquad (4.2.13)$$

where α and β are constants such that the process is stationary, and $\{Z_t\}$ denotes a strict white noise sequence with zero mean and variance σ_Z^2. Now Z_t is independent of X_s for $s < t$, and so we have such general results as $E[Z_t X_s | \text{Data to time } N] = 0$ for $t > s > N$ and $E[Z_t X_s | \text{Data to time } N] = \sigma_Z^2$ for $t = s > N$. The latter result follows because we may write X_t as Z_t plus terms which are independent of Z_t. In order to evaluate the conditional expectation of $Z_t X_s$ when $t < s$, we would have to write X_s in terms of $X_{s-1}, X_{s-2}, \ldots, Z_s, Z_{s-1}, \ldots$ and then iteratively repeat this process until the subscripts of the X-terms are $\leq t$. Applying these results to the simple model above, we have

$$\hat{x}_N(1) = \alpha X_N + \beta Z_N X_N$$

and

$$\hat{x}_N(2) = \alpha \hat{x}_N(1) + \beta \sigma_Z^2$$

Forecasts for longer lead times may be similarly obtained. Depending on the structure of the model, this may be relatively straightforward or algebraically messy.

When analytic point forecasts are difficult, or even impossible, to evaluate, some sort of numerical evaluation is usually needed. This may involve numerical integration, bootstrapping or some sort of Monte Carlo method.

As well as being difficult to evaluate conditional expectations more than one step ahead, another complication with most non-linear models is that the predictive distribution is unlikely to be normally distributed, even if the innovations distribution is assumed to be normal. Furthermore, its variance need not increase monotonically with the forecast horizon. Sometimes, the predictive distribution may not even be unimodal. This means that it is not enough to just calculate the point forecast together with the forecast error variance. Rather, the analyst needs to compute the complete predictive distribution. This can be costly in terms of computer time, though packages are becoming available for carrying out the necessary numerical work. For example, Tong (1990) publicizes a package called STAR. Clements and Smith (1997) describe a simulation study to compare several different ways of computing forecasts for a SETAR model, but no clear picture emerges.

4.3 Ad hoc forecasting methods

This section introduces a number of forecasting methods which are not based explicitly on a probability model, and so might be regarded to some extent as being of an ad hoc nature. However, they will generally be based, at least partly, on a preliminary examination of the data to determine, for example, whether a seasonal or non-seasonal method should be used. Thus they may well depend, at least implicitly, on some sort of model, albeit of a simple form.

4.3.1 Simple exponential smoothing

Perhaps the best known forecasting method is that called *exponential smoothing* (ES). This term is applied generically to a variety of methods that produce forecasts with simple updating formulae and can follow changes in the local level, trend and seasonality. The simplest version, called *simple exponential smoothing* (SES), computes the one-step-ahead forecast by a formula that is equivalent to computing a geometric sum of past observations, namely

$$\hat{x}_N(1) = \alpha x_N + \alpha(1-\alpha)x_{N-1} + \alpha(1-\alpha)^2 x_{N-2} + \dots \qquad (4.3.1)$$

where α denotes the *smoothing parameter* in the range (0,1). In practice, the above equation is always rewritten in one of two equivalent *updating* formats, either the *recurrence* form

$$\hat{x}_N(1) = \alpha x_N + (1-\alpha)\hat{x}_{N-1}(1) \qquad (4.3.2)$$

(which is the same as (4.2.6)) or the *error-correction* form

$$\hat{x}_N(1) = \hat{x}_{N-1}(1) + \alpha e_N \qquad (4.3.3)$$

where $e_N = x_N - \hat{x}_{N-1}(1) =$ prediction error at time N.

The original motivation for SES (Brown, 1963) came from applying *discounted least squares* to a model with a constant local mean, μ, and white noise errors, $\{\varepsilon_t\}$, such that

$$X_t = \mu + \varepsilon_t \qquad (4.3.4)$$

where μ was estimated, not by least squares, but by applying geometrically decaying weights to the squared prediction error terms. In other words, we estimate μ by minimizing

$$S = \sum_{j=0}^{N-1} \beta^j (x_{N-j} - \mu)^2 \qquad (4.3.5)$$

It turns out that the resulting estimate of μ using data up to time N is such that it can be updated by SES (e.g. Kendall and Ord, 1990, Section 8.19).

The above procedure is intuitively reasonable but theoretically dubious in that if one really believes the underlying mean is constant, then ordinary least squares should be used. This implies that the analyst really has in mind a model where the level is locally constant but may change through time, and SES is sometimes said to be applicable for series showing no seasonal variation or long-term trend but with a locally constant mean which shows some 'drift' over time, whatever that means. In fact, it can be shown that SES is optimal for several underlying models where the form of the 'drift' is mathematically specified. One such model is the ARIMA$(0,1,1)$ model given by

$$(1-B)X_t = Z_t + (\alpha - 1)Z_{t-1} \qquad (4.3.6)$$

where the coefficient of ε_{t-1} is expressed in the form $(\alpha - 1)$, rather than as the more usual θ, to clarify the connection with SES having smoothing parameter α. Then it was shown in Section 4.2.1 that the optimal one-step-ahead prediction, namely the conditional mean of the next observation, for the model in (4.3.6) is updated by SES as in (4.3.2) with smoothing parameter α. The invertibility condition for the ARIMA$(0, 1, 1)$ model in (4.3.6) leads to the restriction $0 < \alpha < 2$ on the smoothing parameter. This is wider than the more usual range of $0 < \alpha < 1$. When α takes a value in the range $(1, 2)$, then (4.3.1) shows that the weights attached to past values when producing a forecast will oscillate in sign. This is rather unusual in practice, though it could occasionally be appropriate, for example, when there is 'over-adjustment' to share prices, or when sales are mis-recorded by being credited to the one sales period at the expense of the previous period. Note that when $\alpha = 1$, (4.3.6) reduces to the simple random walk for which $\hat{x}_N(1) = x_N$.

The second model, for which SES is well known to be optimal, is the random walk plus noise state-space model (see Section 2.5.5). When the Kalman filter (see Section 4.2.3) is applied to this model, it can be shown that SES results in the steady state with a smoothing parameter, α, which is a function of the signal-to-noise ratio. If we denote the latter by $c = \sigma_w^2 / \sigma_n^2$, then we find $\alpha = \frac{1}{2}[(c^2 + 4c)^{0.5} - c)]$, Thus exponential smoothing can be regarded as a (very simple) Kalman filter. It can be shown that α is restricted to the range $[0,1]$, essentially because the two error variances must both be non-negative so that c is non-negative.

The ARIMA$(0, 1, 1)$ and 'random walk plus noise' models are both non-stationary, but have first differences which are stationary with an autocorrelation function of the same form, namely a non-zero coefficient at lag one only. Thus the two models have the same second-order structure and it is generally impossible to decide which is the 'better' model for a given set of observed data. Despite this, the two models are quite different in character. The ARIMA model is constructed from a single white noise process, while the 'random walk plus noise' model is constructed from two independent error processes. The invertible ARIMA$(0, 1, 1)$ model allows a smoothing parameter in the range $(0, 2)$, while the random walk plus noise model restricts α to the range $(0, 1)$. So is the ARIMA model more general or is the state-space model more realistic? The answer to this question depends partly on the *context*. For example, if weights with alternating signs in (4.3.1) do not make sense, then it seems unwise to allow α to exceed unity. We say more about this in Section 4.4.

In fact, several other models can be found (Chatfield et al., 2001) for which exponential smoothing gives MMSE predictions, including a state-space model with a non-constant variance. This helps to explain why methods based on exponential smoothing seem to give such robust predictions and suggests that further research is needed to find ways of identifying an appropriate underlying model for such smoothing methods. After all, if exponential smoothing is optimal for models with both constant and non-constant variance, then it will, for example, be important to know which is the more plausible assumption when computing prediction intervals even if the point forecasts are unaffected.

The smoothing parameter, α, is normally constrained to the range $(0,1)$ and is often estimated by minimizing the sum of squared one-step-ahead 'forecast' errors over the period of fit (so that the 'forecasts' are really within-sample fits). The starting value for $\hat{x}_1(1)$ is typically taken to equal x_1. More details are given, for example, by Gardner (1985). The method is now available in many software packages, and is very easy to implement. However, it is inappropriate for series showing trend and seasonality and should therefore be seen primarily as a building block for more general versions of ES, which *can* cope with trend and seasonal variation.

Before introducing these more general versions, it will be helpful to recast SES in the following way, by regarding it as being a method for producing an estimate of the local level at time t, say L_t. Then (4.3.2), for example, is rewritten as

$$L_N = \alpha x_N + (1 - \alpha)L_{N-1} \tag{4.3.7}$$

The one-step-ahead forecast is then given by

$$\hat{x}_N(1) = L_N \tag{4.3.8}$$

and the h-steps-ahead forecast, $\hat{x}_N(h)$, is also equal to L_N.

4.3.2 Holt's linear trend method

The first generalization is to include a local trend term, say T_t, which measures the expected increase or decrease per unit time period in the local mean level. This term is also updated using an equation akin to ES, albeit with a different smoothing parameter, say γ. The updating equation for the local level, L_t, is obtained by generalizing (4.3.7) in an obvious way to give

$$L_N = \alpha x_N + (1 - \alpha)(L_{N-1} + T_{N-1}) \tag{4.3.9}$$

In addition, we update the local estimate of the growth rate by

$$T_N = \gamma(L_N - L_{N-1}) + (1 - \gamma)T_{N-1} \tag{4.3.10}$$

The h-steps-ahead forecast is then given by

$$\hat{x}_N(h) = L_N + hT_N \tag{4.3.11}$$

Equations (4.3.9) and (4.3.10) correspond to the recurrence form of ES in (4.3.2). Alternative formulae, which correspond to the error-correction form of ES in (4.3.3) are also available to update L_N and T_N, though they make the smoothing parameters look different. For example, some simple algebra shows that the formula for updating the growth rate, T_N, in (4.3.10) is given by

$$T_N = T_{N-1} + \alpha\gamma e_N \tag{4.3.12}$$

where it looks as though the smoothing parameter has changed to $\alpha\gamma$.

The above procedure is described as being appropriate for series showing a 'local linear trend' and is usually known as *Holt's linear trend procedure*. It can be shown that the method is optimal when the underlying process is either an ARIMA$(0, 2, 2)$ model or a linear growth state-space model. There are now two smoothing parameters to estimate and starting values for both the level and trend must be provided. The latter are usually calculated from the first year or two's data in a simple-minded way.

A useful variant of Holt's linear trend method is to include a *damping parameter*, say ϕ, where $0 < \phi < 1$, such that the estimate of the trend or growth rate at time t, say T_t, is damped to ϕT_t in the subsequent time period. The h-steps-ahead forecast at time N is then

$$x_N(h) \;=\; L_N \;+\; (\sum_{i=1}^{h} \phi^i) T_N \qquad (4.3.13)$$

This method involves estimating one more parameter, ϕ, and is optimal for a particular ARIMA$(1, 1, 2)$ model. Further details may be found in Gardner and McKenzie (1985) who show that the method often gives good empirical results.

There is a technique called *double exponential smoothing* which is a special case of Holt's linear trend method (and also of general exponential smoothing – see Section 4.3.4). As the name implied, SES is used to update the local estimate of level and then the same operation, with the same smoothing parameter, is applied to the sequence of estimates of level. It can be shown that this method arises when the underlying ARIMA$(0, 2, 2)$ model has equal roots. When the double exponential smoothing equations are rewritten in the Holt form, the two Holt smoothing parameters are related to each other but in a less obvious way – Gardner (1985) gives the necessary formulae. Thus this method effectively reduces the number of smoothing parameters to one. There seems little point in making this restriction and the method is seldom used today.

4.3.3 The Holt-Winters forecasting procedure

Exponential smoothing can further be generalized to cope with seasonal variation of either an *additive* or *multiplicative* form. Let I_t denote the *seasonal index* in time period t, s the number of periods in one year (e.g. $s = 4$ for quarterly data), and δ the smoothing parameter for updating the seasonal indices. In the additive case, the seasonal indices are constrained so that they *sum* to zero over a full year, while in the multiplicative case they should *average* to unity. Taking the additive case as an example, the recurrence form of the updating equations for the level, L_N, the growth rate, T_N, and the seasonal index, I_N, are

$$L_N = \alpha(x_N - I_{N-s}) + (1 - \alpha)(L_{N-1} + T_{N-1}) \qquad (4.3.14)$$

$$T_N = \gamma(L_N - L_{N-1}) + (1 - \gamma)T_{N-1} \qquad (4.3.15)$$

$$I_N = \delta(x_N - L_N) + (1 - \delta)I_{N-s} \qquad (4.3.16)$$

The forecasts at time N are then given by

$$\hat{x}_N(h) = L_N + hT_N + I_{N-s+h} \qquad (4.3.17)$$

for $h = 1, 2, \ldots, s$. Alternative error-correction formulae are also available
to update L_N, T_N, and I_N, though, as for Holt's linear trend method, they
make the smoothing parameters appear to be different.

The seasonal form of exponential smoothing is usually called *Holt-Winters seasonal forecasting*. It can be shown that the version assuming
additive seasonality is optimal for a seasonal ARIMA model of a rather
complicated form which would probably never be identified in practice.
The multiplicative version has a non-linear form and so has no ARIMA
representation (as ARIMA models are inherently linear). However, recent
work by Ord et al. (1997) on a class of non-linear state space models has
led to several models for which the optimal set of updating equations is
very close to the usual multiplicative Holt-Winters formulae. This will help
to provide some statistical foundation to the method.

A graph of the data should be examined to see if seasonality is present,
and, if so, whether it is additive or multiplicative. The user must provide
starting values for L_t, T_t and I_t at the start of the series, estimate the three
smoothing parameters over some suitable fitting period, and then produce
forecasts from the end of the series. The method is straightforward and
widely used. It can be implemented in an automatic way for use with a
large number of series or can be implemented in a non-automatic way
with, for example, a careful assessment of outliers. Empirical results (see
Chapter 6) suggest that the method has accuracy comparable to that of
other seasonal procedures.

For further details and practical guidance on implementing the different
versions of ES, the reader is referred to Gardner (1985), Chatfield and Yar
(1988) and Chatfield (1996a).

4.3.4 Other methods

This subsection briefly introduces several more forecasting methods of a
generally ad hoc nature.

(i) *Stepwise autoregression.* This automatic procedure can be regarded as
a subset of the Box-Jenkins procedure in that first differences are taken
of the series of observations to remove (most of) any trend, and then a
subset regression package is used to fit an AR model to the first differences.
The lags that are included in the model are chosen by applying a forward
selection procedure of a similar type to that used in subset selection in

multiple regression. In other words, the analyst starts by fitting the 'best' lagged variable, and then introduces a second lagged variable, provided this leads to a significant reduction in the residual sum of squares, and so on. It is sensible to choose a maximum possible lag of perhaps twice the number of observations in a season. Thus the analyst could go up to a maximum lag of say 8 or 9 periods for quarterly data. The rationale behind the method is that AR models are much easier to fit than MA models, and the analyst effectively ends up with an integrated AR model with $d = 1$. Further details are given in Granger and Newbold (1986, Section 5.4). Now that the full Box-Jenkins procedure is much easier to implement, the method is rarely used.

(ii) *The ARARMA method.* This method, proposed by Parzen (1982) and also described in Makridakis et al. (1984), relies on fitting a possibly non-stationary AR model to the data to remove any trend, before fitting an ARMA model to the detrended data. This explains the title of the method. If, for example, an AR(1) model is fitted to data which exhibits trend, then the fitted AR parameter will typically be close to one (Parzen calls it a long-memory model if the estimated coefficient exceeds 0.9) or could even exceed one (a non-stationary value). An empirical study by Meade and Smith (1985) suggests that the benefits of the ARARMA approach may lie in the increased emphasis on correctly identifying the form of the trend, allied to a reluctance to over-difference, rather than because the AR transformation to stationarity is inherently superior to differencing. Brockwell and Davis (1996, Section 9.1.1) describe a variant of this procedure in which no MA terms are fitted so that the procedure is called the ARAR algorithm.

(iii) *General exponential smoothing (GES).* This method was proposed by Brown (1963) and was very influential in its day. More recent accessible summaries are included in Gardner's (1985) review and in Abraham and Ledolter (1986). Suppose we can represent a time series in terms of standard functions of time such as polynomials and sinusoids, in the form

$$X_t = \sum_{i=1}^{k} a_i f_i(t) \; + \; \varepsilon_t \qquad (4.3.18)$$

where $\{a_i\}$ denote a set of constants and $f_i(t)$ is a known function of time. Although the latter are deterministic functions, the method recognizes that the coefficients $\{a_i\}$ will change through time, so that the model only applies locally. Thus estimates of the $\{a_i\}$ are updated through time as more data become available. As the model only holds locally, it is natural to estimate the parameters at time t, not by least squares, but by giving more weight to recent observations. This suggests using *discounted least squares*, wherein the analyst minimizes the discounted sum of squared errors at time

t, namely

$$S = \sum_{j=0}^{\infty} \beta^j (x_{t-j} - \sum_{i=1}^{k} a_i(t) f_i(t - j))^2 \qquad (4.3.19)$$

where we now write the $\{a_i\}$ as functions of t to show that they are local parameters. The estimating equations that result give estimates of the $\{a_i(t)\}$, say $\{\hat{a}_i(t)\}$, which can usually be updated in a fairly straightforward way using the latest observation and the most recent parameter estimates, namely $\{\hat{a}_i(t-1)\}$.

In practice, the GES model is usually formulated so that the $\{f_i\}$ are functions of the time difference from the end of the series rather than from the origin. For example, when there are $k = 2$ functions, with f_1 a constant and f_2 a linear trend, then we want the forecast function to be of the form

$$\hat{x}_t(h) = \hat{a}_1(t) + \hat{a}_2(t)h \qquad (4.3.20)$$

where $\hat{a}_1(t)$, $\hat{a}_2(t)$ denote estimates of the level and trend at time t, and so we would replace the sum of squares in (4.3.19) by

$$S = \sum_{j=0}^{\infty} \beta^j (x_{t-j} - a_1(t) - a_2(t)j)^2 \qquad (4.3.21)$$

When the only function fitted is a constant, then it can readily be shown that GES reduces to simple exponential smoothing. When a constant and a linear trend is fitted, then it can be shown that GES reduces to double exponential smoothing, which is a special case of Holt's linear trend method as noted in Section 4.3.2. Thus Gardner (1985) concludes that, if data are non-seasonal, then there is little point in using GES.

If data are seasonal, then GES is very different from Holt-Winters. The latter fits one seasonal index for each period, while GES fits several sine and cosine terms. With s periods per year, no more than $(s - 1)$ sinusoidal terms are needed and it may be possible to use fewer terms. GES has one smoothing parameter, while Holt-Winters has three. Is it reasonable to expect one smoothing parameter to do all the work for which Holt-Winters requires three parameters?

Several variants of seasonal GES have been proposed, incorporating more than one smoothing parameter and alternative ways of representing seasonal variation. Examples include the methods called SEATREND and DOUBTS in Harrison (1965). Various empirical studies are reviewed by Gardner (1985, Section 4.7) and my impression is that there is rather little to choose between these methods, provided they are implemented in a sensible way.

(iv) *Others.*   Several other univariate time-series methods have been proposed over the years, such as AEP, adaptive filtering, and the FORSYS method, but are not now generally in use, either because of theoretical problems, or because it is unclear how to implement the method in practice, or because empirical evidence suggests that other methods are better anyway. Thus these methods will not be covered here.

4.3.5 Combining forecasts

In some situations, the analyst may have more than one possible forecast. Rather than choose one of these as somehow 'best', the analyst may choose to combine the forecasts in some way, perhaps by taking the average of the different forecasts. This idea goes back many years, at least to Bates and Granger (1969). A comprehensive review is given by Clemen (1989) in a special section on *Combining Forecasts* in the *Int. J. of Forecasting*, 1989, No. 4. Another review is given by Granger (1989). The recent text by Diebold (1998, Section 12.3) also covers some of the material. Empirical results are very encouraging in that the average of several forecasts typically gives better forecasts than the individual forecasts.

A combined forecast is generally a weighted average of two or more forecasts. For simplicity, consider the case of two forecasts, say f_1 and f_2. A simple linear combination of these quantities produces a new forecast, say

$$f_3 = \lambda f_1 + (1 - \lambda) f_2 \qquad (4.3.22)$$

where λ is usually restricted to the range $(0, 1)$. If the entire weight is placed on one forecast, then that forecast is said to *encompass* the other forecast(s). For example, for the simple linear combination in (4.3.22), f_1 encompasses f_2 if $\lambda = 1$. The value of λ can sometimes be chosen by theory. For example, if f_1 and f_2 are both unbiased with respective variances v_1 and v_2, then, under certain assumptions, it can be shown that the 'optimal' unbiased forecast (meaning the unbiased forecast with minimum variance) is given by (4.3.22) with $\lambda = v_2/(v_1 + v_2)$. However, in practice we are unlikely to know if the forecasts really are unbiased, while the true variances will have to be estimated. Moreover, different forecasts are likely to be correlated. Thus it is more common to simply adopt the pragmatic approach of taking the average. It is for this reason that the approach is included in Section 4.3.

Although good point forecasts may result, a drawback to combining forecasts is that the analyst does not end up with a single model to interpret, so as to better understand the data. Moreover, there is no obvious theoretical way to compute prediction intervals, though alternative empirical methods can be found (Taylor and Bunn, 1999). Thus, although some combination of forecasts may serve the immediate purpose of getting a forecast which is better than the individual forecasts, Clements and Hendry (1998a, Chapter 10) suggest this is nothing more than a convenient stop-gap. The fact that forecasts need to be combined suggests that all the models are mis-specified and so it may be better to devote more effort to refining the best of the models so that it encompasses the rest, or to find a more theoretically satisfying way of combining several plausible models, such as Bayesian Model Averaging – see Section 7.5.7.

4.3.6 Monitoring ad hoc forecasts

Whatever forecasting method is adopted, it is essential to continually check the forecasts so that, if things go wrong, the method or model can be revised as necessary. In Section 3.5.3, we described how the one-step-ahead forecast errors can be used to carry out diagnostic checks on a model, and can also be used for forecast monitoring by plotting them one-at-a-time as each new observation becomes available.

For ad hoc methods, there is no model to check, but it is still sensible to look at the in-sample forecast errors to make sure that the method is performing in a reasonably sensible way. The out-of-sample forecast errors can then be used for forecast monitoring in much the same way as for a model-based method. Is the mean forecast error about zero over some suitable interval? If not, the forecasts are biased. Are the forecast errors autocorrelated? If they are, then the method is not optimal. Are the errors normally distributed? If not, then prediction intervals will not be symmetric. Any indication that the forecast errors have unexpected properties may lead to appropriate corrective action. Devices like tracking signals (see Section 3.5.3) may help.

4.4 Some interrelationships and combinations

This section explores some of the many interrelationships between different linear models and between different forecasting methods in the univariate case. Methods which involve fitting a probability model to the data, and then finding optimal forecasts conditional on that model, are quite different in spirit to forecasts which are not based on a model. Even so, the different approaches may, on occasion, give identical forecasts. Model-based methods are customarily used in statistical forecasting exercises involving a small number of series, while operational research problems, such as inventory control with many items, are customarily tackled with automatic methods which are not model-based.

To illustrate the different approaches, consider exponential smoothing. This is optimal for an $ARIMA(0, 1, 1)$ model or a 'random walk plus noise' model, but is sometimes used in practice for series showing no obvious trend or seasonality without any formal model identification. In other words it can be used, either as the outcome of a model-based approach, or in an ad hoc way.

More generally, the Box-Jenkins forecasting procedure involves formulating, fitting and then checking an appropriate ARIMA model. The corresponding optimal forecasts from that model can then be computed. This approach may be appropriate with only a few series to forecast and with adequate statistical expertise available. At the other extreme, the practitioner with a large number of series to forecast, may decide to use the same all-purpose procedure whatever the individual series look like. For example, the Holt-Winters forecasting procedure may be used for a group of series showing trend and seasonal variation. As no formal model identification is involved, it is inappropriate to talk about the Holt-Winters *model*, as some authors have done.

We concentrate on relationships between models, while making links to methods where appropriate. First we note that interrelationships between models need not necessarily be helpful. For example, it is possible to represent some growth curve models in state-space form or to regard them as a type of non-linear model. However, they are usually best considered simply in their original growth curve form. Thus we concentrate here on relationships which will hopefully help to improve understanding.

First we look at state-space models. In Section 3.2, we pointed out that there are several classes of model which are all essentially of this form. They include the dynamic linear models of West and Harrison (1997), the structural models of Harvey (1989) and the unobserved component models used primarily by econometricians (e.g. Nerlove et al., 1979). Furthermore, there are also close links between state-space models and ARIMA models in that an ARIMA model can generally be expressed in state-space form, though not necessarily in a unique way. For example, in Section 3.2, we showed that an AR(2) model can be expressed as a state-space model in several ways with state vectors involving the last two observations. It is doubtful if this rather contrived piece of mathematical trickery improves our understanding of the model, but it does replace two-stage dependence with one-stage dependence and allows the Kalman filter to be used to update model parameter estimates.

In order to explore the link between ARIMA and state-space models further, we reconsider the two models for which exponential smoothing is well known to be optimal, namely the ARIMA$(0, 1, 1)$ model and the random walk plus noise model. As noted in Section 4.3.1, both models are non-stationary, but can be made stationary by first differencing when they both yield ac.f.s having a non-zero value at lag one only. The parameter θ of the ARIMA$(0, 1, 1)$ model, expressed in the form $X_t - X_{t-1} = Z_t + \theta Z_{t-1}$, must lie within the range (-1,+1) for invertibility, so that the smoothing parameter $\alpha = (1+\theta)$ will lie in the range (0,2). However, the parameters of the random walk plus noise model are constrained so that $0 < \alpha < 1$. This restriction can be seen as good (more realistic) or bad (more restrictive). When the smoothing parameter is larger than unity, the weights attached to past values when making a forecast will oscillate in sign. This would not usually make sense from a practical point of view and it is arguable that the 'random walk plus noise' model is intuitively more helpful in providing an explicit model of changes in the local level. Thus the the use of the state-space model seems generally more helpful and prudent.

Moving on to the seasonal case, the links between seasonal ARIMA models and seasonal state-space models are rather more tenuous. We use the seasonal version of exponential smoothing, namely the Holt-Winters method, to explore these links. It can be shown (Abraham and Ledolter, 1986) that the additive version of Holt-Winters is optimal for a SARIMA model having one non-seasonal and one seasonal difference and with moving average terms at all lags between one and $(s + 1)$. The latter coefficients can all be expressed in terms of the three Holt-Winters smoothing parameters for adjusting the level, trend and seasonal index, respectively. This SARIMA model is so complicated that it is unlikely that it would ever be identified in practice. As to the relationship with structural (state-space) models, it can be argued that the use of the Basic Structural Model (BSM – see Section 3.2) is similar in spirit to the results produced by the additive Holt-Winters method. The latter depends on three smoothing parameters which correspond in some sense to the three error variance ratios of the BSM, namely σ_1^2/σ_n^2, σ_2^2/σ_n^2, and σ_3^2/σ_n^2 in an obvious notation. However, there is not an exact equivalence. Thus the three forecasting methods, namely Box-Jenkins, Holt-Winters and structural modelling, are different to a greater or lesser extent. The Holt-Winters method is clearly the easiest to use and understand, but the use of a proper probability model, such as an ARIMA or structural model, means, for example, that it is easier to make valid probability statements about forecasts, while models are easier to extend to incorporate explanatory variables.

The above remarks help to explore the relationship between ad hoc forecasting *methods* and time-series *models*, in that most examples of the former are optimal for a particular model. Thus, although no formal model identification is typically carried out, it is customary to employ some simple descriptive tools to ensure that the method is at least not unreasonable. For example, a check of the time plot will indicate the possible presence of trend and seasonality and hence indicate which type of exponential smoothing to use. Thus, there is less of a clearcut distinction between ad hoc and model-based forecasting methods than might be thought.

A more difficult question is whether, and when, it is reasonable to use the properties of the implicit 'optimal' model even when no formal identification has been made. For example, forecasters often use results relating to the ARIMA$(0, 1, 1)$ model when using simple exponential smoothing. This is not strictly valid but is arguably reasonable provided that some preliminary assessment of the data has been made to ensure the method is plausible and that checks on the forecasts also indicate no problems.

Linked to the above is the question as to whether exponential smoothing should be regarded as a special case of the Box-Jenkins method as has been suggested elsewhere (e.g. Jenkins, 1974). My view is that it should *not* be so regarded. The way that the methods are implemented in practice is quite different even if some exponential smoothing methods are optimal for some special cases of the ARIMA class. The way that the model or method is identified, the way that starting values are chosen and the way that parameters are estimated are quite different. Moreover, non-ARIMA models, including models with changing variance, can be found for which exponential smoothing methods are optimal (see Section 4.3.1), so that the latter are, in some ways, *more* general than Box-Jenkins. As regards seasonal exponential smoothing, it was noted earlier that, although the additive version of Holt-Winter is optimal for a complicated SARIMA model, it is unlikely that such a model would ever be identified in practice. Moreover, the multiplicative version of Holt-Winters does not correspond to any ARIMA model. Thus, for all practical purposes, Holt-Winters is not a special case of the Box-Jenkins method.

Turning now to the *combination* of methods and models, Section 4.3.5 pointed out that better forecasts can often be obtained by taking a weighted average of forecasts from different methods. This implies that none of the methods is optimal by itself, which ties in with the suggestion that there is usually no such thing as a 'true' model. A drawback to combining forecasts is that no single model applied. This makes it potentially attractive to consider ways in which models may be combined in order to get a better approximation to reality with a single model. There are, of course, many different ways in which this can be done as, for example, combining regression models with ARCH disturbances. However, it is not always straightforward to combine models as Example 4.1. is meant to illustrate.

Example 4.1 Suppose we postulate a model which has systematic variation following a deterministic linear trend with time, but with innovations that are not independent, but rather follow an AR(1) process. In other words, we combine a linear trend with autocorrelated errors, and our model is

$$X_t = \alpha + \beta t + u_t \qquad (4.4.1)$$

with

$$u_t = \phi u_{t-1} + \varepsilon_t \qquad (4.4.2)$$

where α, β and ϕ are constants and ε_t denotes a purely random process. With some simple algebra, this can be rewritten in reduced form as the single equation

$$X_t = \gamma + \delta t + \phi X_{t-1} + \varepsilon_t \qquad (4.4.3)$$

where $\gamma = \alpha(1 - \phi) + \beta\phi$, and $\delta = \beta(1 - \phi)$. When $\phi = 1$, it is easy to see that $\delta = 0$ and the equation has a unit root, so that the model is difference-stationary (see Section 3.1.9). However, if $\phi < 1$ and $\beta \neq 0$, then the model is trend-stationary. (Of course, if $\phi < 1$ and $\beta = 0$, then X_t is stationary anyway.) When $\beta \neq 0$, an interesting feature of the revised formulation is that the coefficients of t in (4.4.1) and (4.4.3) are not the same, since $\delta \neq \beta$, and so the slope may appear to be different when expressed as in (4.4.3). As it is non-stationary (for $\beta \neq 0$), it seems natural to take first differences, but the resulting model will not be invertible, and is therefore not identifiable, when $\phi < 1$. Thus this innocent-looking model is actually rather hard to handle and there can be problems fitting it.   □

A moral of Example 4.1 is that great care is needed when combining models of different types as their properties may prove incompatible. If you want to model a deterministic linear trend, it is unwise to add innovations which follow an AR(1) process. More generally, this section demonstrates that many univariate models and methods are interlinked, that studying these links may lead to greater insight, but that ad hoc methods nominally optimal for a particular model should not be confused with a model-based method.

CHAPTER 5

Multivariate Forecasting Methods

"Forecasting is the art of saying what will happen and then explaining why it didn't!"
— Anonymous
"An economist is an expert who will know tomorrow why the things he predicted yesterday didn't happen today."
— Evan Esar

5.1 Introduction

Observations are often taken simultaneously on two or more time series. For example, we might observe various measures of economic activity in a particular country at regular intervals of time, say monthly. The variables might include the retail price index, the level of unemployment and a weighted index of share prices. Given such multivariate data, it may be desirable to try to develop a multivariate model to describe the interrelationships among the series, and then to use this model to make forecasts. With time-series data, the modelling process is complicated by the need to model, not only the interdependence *between* the series, but also the serial dependence *within* the component series.

This chapter introduces a variety of multivariate time-series models and the forecasting methods which are based on them. The models include multiple regression, transfer function and distributed lag models, econometric models and multivariate versions of AR and ARMA models, including vector autoregressive (VAR) models. Alternative introductions are given by Granger and Newbold (1986, Chapters 7 and 8), Priestley (1981, Chapter 9) and Wei (1990, Chapters 13 and 14).

Fitting multivariate models to time-series data is still not easy despite enormous improvements in computer software in recent years. There is much ongoing research, partly stimulated by improved computational resources, but much remains to be done.

The presentation of material in this chapter differs from that on univariate methods which was presented over two chapters, one on models (Chapter 3) and a second on methods (Chapter 4). Here the emphasis is primarily on models. In particular, there is no section analogous to Section 4.3, which discussed univariate ad hoc forecasting methods not based explicitly on a model. This is because there are essentially no ad hoc multivariate methods. Furthermore there is no need to re-present the general material on prediction from Section 4.1 as the comments therein will also apply to multivariate forecasting with obvious amendments. In particular, we assume Equation (4.2.1) throughout, namely that, having identified a model, M say, the optimum forecast of X_{N+h} at time N is obtained by finding $E[X_{N+h}|M, I_N]$ where I_N denotes the information available at time N. However, there is one important difference in that this latter expression may need to incorporate forecasts of explanatory variables. As we see below, this may make forecasts of the dependent variable less accurate than might be expected.

While univariate models can be useful for many purposes, including forecasting large numbers of series, and providing a benchmark in comparative forecasting studies, it seems clear that multivariate models should also have much to offer in gaining a better understanding of the underlying structure of a given system and (hopefully) in getting better forecasts. Sadly, we will see that the latter does not always happen. While multivariate models can usually be found which give a better *fit* than univariate models, there are several reasons why better forecasts need not necessarily result (though of course they sometimes do). The reasons include the following (in roughly ascending order of importance):

(i) With more parameters to estimate, there are more opportunities for sampling variation to increase parameter uncertainty and affect forecasts.

(ii) With more variables to measure, there are more opportunities for errors and outliers to creep in.

(iii) Observed multivariate data may not necessarily be suitable for fitting a multivariate model – see Section 5.1.1 below.

(iv) The computation of forecasts of a dependent variable may require future values of explanatory variables which are not available at the time the forecast is to be made. Then the explanatory variables must be predicted in some way before forecasts of the dependent variable can be found and this inevitably leads to a reduction in accuracy. If forecasts of explanatory variables have poor accuracy, then the resulting forecasts of the dependent variable may have worse accuracy than univariate forecasts (Ashley, 1988).

(v) The computation of multivariate forecasts depends on having a good multivariate model, but this cannot be guaranteed. As for univariate models, a multivariate model may be incorrectly identified or may change over the period of fit or in the future. It appears that multivariate models, being more complicated, are more vulnerable to misspecification than univariate models. Moreover, multivariate modelling is generally much more difficult to carry out than fitting univariate models.

Experience in searching for an appropriate multivariate model substantiates the importance of getting sufficient background information so as to understand the context and identify all relevant explanatory variables. This may not be easy and, as always, it is vital to ask lots of questions and see, for example, if there are any previously known empirical relationships between the measured variables. It is also vital to formulate the problem carefully. An iterative approach to model building (see Sections 3.5, 8.2 and 8.4) is generally required, and the use to which the model will be put (e.g. forecasting) should be considered as well as the goodness-of-fit. There is always tension between including unnecessary explanatory variables, which appear to improve the fit but actually lead to worse out-of-sample forecasts, and omitting crucial variables which are needed. Put another way, it is desirable to seek a parsimonious model (so that fewer parameters need to be estimated) while ensuring that important variables are not ignored. However, whatever model is eventually fitted, statistical inference is normally carried out conditional on the fitted model, under the assumption that the model is 'true'. As a result it focusses on uncertainty due to sampling variation and having to estimate model parameters, even though specification errors (due to using the wrong model) are likely to be more serious. The topic of model uncertainty is explored in depth in Chapter 8.

This introductory section goes on the consider two important questions that arise with respect to multivariate forecasting, namely (i) Is feedback present? and (ii) Are forecasts genuinely out-of-sample? Then Section 5.1.3 defines the cross-covariance and cross-correlation matrix functions for stationary multivariate processes, while Section 5.1.4 describes how to carry out an initial data analysis in the multivariate case. Finally, Section 5.1.5 mentions some difficulties in multivariate modelling.

5.1.1 Is feedback present?

One basic question is whether a multivariate model should involve a single equation or several equations. In a (single) multiple regression equation, for example, the model explains the variation in a *dependent* or *response* variable, say Y, in terms of the variation in one or more *predictor* or *explanatory* variables, say $X_1, X_2, \ldots$. There should be no suggestion that the value of the response variable could in turn affect the predictor variables, giving rise to what is often called *feedback*. In other words the regression equation assumes that the variables constitute what is called an *open-loop* system. Some people would then say that there is a *causal relationship* between the explanatory variables and the response variable, though in practice it is difficult to decide if there is a direct link or if there is an indirect link via additional variables not included in the model.

A completely different situation arises when the 'outputs' affect the 'inputs' so that there is a *closed-loop* system. As a simple example of such a system in economics, it is known that a rise in prices will generally lead to a rise in wages which will in turn lead to a further rise in prices. Then a regression model is not appropriate and could well give misleading results if mistakenly applied. Rather a model with more than one equation will be needed to satisfactorily model the system.

Building a 'good' model from data subject to feedback can be difficult.

A well-controlled physical system, such as a chemical reactor, is typically managed by some sort of feedback control system. Then there may not be enough information available in observed data, taken while the system is 'in control', to successfully identify the structure of the system, and it may be necessary to superimpose known perturbations on the system in order to see what effect they have. In contrast an economy is typically subject to some sort of feedback, either from government decisions and/or from personal choice, but is generally not as well controlled as a physical system such as a chemical reactor. Efforts to control the economy are often made in a subjective way and the amount of information in the system may be less than one would like. Furthermore, it is difficult to carry out experiments on an economy in the same sort of way that perturbations can be added to a physical system. Further comments on economic data are made in Section 5.5.

5.1.2 Are forecasts out-of-sample?

In Section 1.4, the importance of distinguishing between in-sample and out-of-sample forecasts was emphasized. This distinction is fairly clear for univariate forecasting, but is much less clear for multivariate forecasting where various types of forecast are possible depending on the type of model and what is known or assumed about the values of explanatory variables.

Given multivariate data measured up to time N, forecasts of future values of the response variable that only use information up to time N about both the response and explanatory variables, are definitely out-of-sample forecasts (often called *ex-ante* forecasts in an economic context). If forecast formulae for the response variable involve future values of the explanatory variables, then an out-of-sample forecast will require the computation of forecasts of such values, either using judgement or by applying some univariate or multivariate procedure. However, it is helpful to consider the following variants. Suppose the observed data are divided into a training sample of length N_1 and a test sample of length N_2. Typically N_1 is (much) larger than N_2. If a model is fitted to the training sample, an assessment of its forecasting accuracy can be made by 'forecasting' the known values of the response variable in the test period. This is often done by using the known values of the explanatory variables, and this procedure is called *ex-post* forecasting in an economic context. The results can help to assess the suitability of the model for forecasting but the forecast accuracy will appear to be more accurate than would be the case in a true out-of-sample (*ex-ante*) situation where future values of the explanatory variables are unknown.

Some writers (e.g. Granger and Newbold, 1986) distinguish between what are sometimes called conditional and unconditional forecasts (though the reader may not find these terms particularly helpful). If forecasts of a response variable are calculated given known or assumed future values of explanatory variables, then they are said to be *conditional* forecasts. They arise in *ex-post* forecasting, and also in *ex-ante* economic forecasting where a policy maker wants to look at the forecasts of a variable of interest which would result from setting future values of explanatory variables according to various policy decisions, rather than by forecasting them. This latter case is an example of 'What-if' forecasting (sometimes called scenario analysis or contingency analysis). In contrast, *unconditional* forecasts of the response variable are said to be obtained if nothing is known or assumed about the future and any future values of explanatory variables which are required are themselves forecasted. This terminological distinction between conditional and unconditional forecasts seems rather unhelpful given that all forecasts are conditional on information of some sort.

As discussed further in Section 5.2, the search for genuine out-of-sample forecasts, and the desirability of not having to forecast explanatory variables, explains the search for what are called *leading indicators*. Consider a simple regression equation, such as

$$Y_t = a + bX_{t-d} + \varepsilon_t \qquad (5.1.1)$$

where Y_t, X_t denote the response and explanatory variables, respectively, a, b, d are constants, with d a non-negative integer, and $\{\varepsilon_t\}$ denotes a white noise process. If d is an integer greater than zero, then X_t is said to be a *leading indicator* for Y_t, and the model enables forecasts of Y_t to be made for up to d steps ahead without having to forecast the X_t series as well. To forecast more than d steps ahead, the required value of X_t will not be available and must itself be forecasted, but the necessary lead time for forecasting a leading indicator will at least be less than for the response variable. Generally speaking, a multivariate model is able to give 'good' forecasts of the response variable provided that any necessary explanatory variables are known (because they are leading indicators) or can be forecast accurately.

5.1.3 Cross-correlations for stationary multivariate processes

A key tool in modelling multivariate time-series data is the cross-correlation function which is used to help describe models having the property of stationarity. Univariate stationary processes were introduced in Section 2.4. This subsection extends the concept of stationarity to the multivariate case and then defines the cross-covariance and cross-correlation matrix functions.

The cross-covariance function and cross-spectrum for a bivariate stationary process were defined in Section 2.4 and we now generalize the discussion to a set of m processes, say $X_{1t}, X_{2t}, \ldots, X_{mt}$. The vector of values at time t will be denoted by $\boldsymbol{X}_t$ where $\boldsymbol{X}_t^T = (X_{1t}, X_{2t}, \ldots, X_{mt})$. Let $\boldsymbol{\mu}_t$ denote the vector of *mean* values at time t so that its ith component is $\mu_{it} = E(X_{it})$. Let $\Gamma(t, t+k)$ denote the *cross-covariance matrix* between $\boldsymbol{X}_t$ and $\boldsymbol{X}_{t+k}$, so that its (i,j)th element is the cross-covariance coefficient of X_{it} and $X_{j,t+k}$, namely $E[(X_{it} - \mu_{it})(X_{j,t+k} - \mu_{j,t+k})]$ – see Equation

(2.4.4). Then this multivariate process is said to be *second-order stationary*
if the mean and the cross-covariance matrices at different lags do not
depend on time. In particular, $\boldsymbol{\mu}_t$ will be a constant, say $\boldsymbol{\mu}$, while $\Gamma(t, t+k)$
will be a function of the lag, k, only and may be written as $\Gamma(k)$.

In the stationary case, the set of cross-covariance matrices, $\Gamma(k)$ for
$k = 0, \pm1, \pm2, \ldots$, is called the *covariance matrix function*. It appears to
have rather different properties to the (auto)covariance function in that it
is not an even function of lag, but rather we find the (i, j)th element of
$\Gamma(k)$ equals the $(j, i)$the element of $\Gamma(-k)$ so that $\Gamma(k) = \Gamma^T(-k)$ (though
the diagonal terms, which are auto- rather than cross-covariances, do have
the property of being an even function of lag). Given the covariance matrix
function, it is easy to standardize any particular element of any matrix
(by dividing by the product of the standard deviations of the two relevant
series) to find the corresponding cross-correlation and hence construct the
set of $(m \times m)$ cross-correlation matrices, $P(k)$ for $k = 0, \pm1, \pm2, \ldots$, called
the *correlation matrix function* of the process.

5.1.4 Initial data analysis

In Sections 2.3 and 3.5, we stressed the importance of starting any analysis
with an initial examination of the data, and in particular of looking at
the time plot of a given series. That recommendation is now extended to
multivariate data. With two or more time series, it is essential to begin
by plotting each variable and looking carefully at the resulting graphs.
As indicated in Section 2.3.1, the time plots should reveal the presence of
trend, seasonality, outliers and discontinuities in some or all of the series.

As a result of looking at the data, it may be evident that some data
cleaning is necessary. In Section 2.3.4, we saw that dealing with outliers,
missing observations and other peculiarities can be difficult and that it is
hard to give general advice on such context-dependent problems. It is,
however, worth reiterating the general advice, from Section 2.3.7, that
the choice of action on such matters can be at least as important as
the choice of forecasting model. As well as cleaning the data, it may be
appropriate to transform the data so as to remove trend and/or seasonality
and/or changing variance, perhaps by applying differencing and/or a power
transformation. This important topic is discussed elsewhere, notably in
Section 2.3.3.

In order to look at the interrelationships *between* series, it can be helpful
to construct some alternative, more complicated, graphs, such as plotting
more than one series on the same time plot. However, we defer discussion of
such graphs to later in this subsection after we have introduced the sample
cross-correlation function.

For series which appear to be (at least approximately) stationary, it is
usually helpful to calculate the *sample mean vector*, the *autocorrelation
function* for each series, and, in addition, the *sample cross-correlation
function* for all meaningful pairs of variables. Suppose we have N pairs of

observations $\{(x_t, y_t), t = 1, \ldots, N\}$ on two series labelled x and y such that x_t and y_t are measured at the same time, t. Then the sample mean vector, denoted by $(\bar{x}, \bar{y})$, provides an estimate of the population mean vector, say (μ_x, μ_y), on the assumption that the bivariate process is stationary (and ergodic). The sample cross-covariance function is given by

$$c_{xy}(k) = \begin{cases} \sum_{t=1}^{N-k}(x_t - \bar{x})(y_{t+k} - \bar{y})/N & k = 0, 1, \ldots, N-1 \\ \sum_{t=1-k}^{N}(x_t - \bar{x})(y_{t+k} - \bar{y})/N & k = -1, -2, \ldots, -(N-1) \end{cases}$$
(5.1.2)

Assuming stationarity, the coefficient for lag k, namely $c_{xy}(k)$, provides an estimate of the population cross-covariance coefficient, $\gamma_{xy}(k)$, defined in (2.4.4) or equivalently the (x, y)-th element of the population cross-covariance matrix, $\Gamma(k)$, defined in Section 5.1.3.

The sample cross-correlation function is given by

$$r_{xy}(k) = c_{xy}(k)/\sqrt{[c_{xx}(0)c_{yy}(0)]}$$
(5.1.3)

where $c_{xx}(0)$ and $c_{yy}(0)$ are the sample variances of the observations on the x- and y-series, respectively. For stationary series, the sample cross-correlation function provides an estimate of the theoretical cross-correlation function defined in Equation (2.4.5). Unlike the autocorrelation function, the cross-correlation function is not an even function of lag and the value of the lag which gives the maximum cross-correlation may help to indicate which series is 'leading' the other.

For appropriate data, the size and position of the maximum cross-correlation may be used to indicate the strength and direction of the relationship between the two series. But what is meant by 'appropriate data'? In practice, the sample cross-correlation function is notoriously difficult to interpret, as illustrated in Section 5.2.1 with a discussion of the famous example in Box and Newbold (1971). In brief, cross-correlations *between* series are affected by serial dependence (autocorrelation) *within* the component series and so may, on occasion, present misleading information. This statement depends on the following results.

In the univariate case, it is easy to test whether an *auto*correlation coefficient is significantly different from zero, by seeing whether it exceeds the bounds $\pm 2/\sqrt{N}$. Unfortunately, the corresponding test for cross-correlations is much more difficult because the formula for the variance of $r_{xy}(k)$, assuming the series are not cross-correlated, involves the autocorrelations of both series, and is given by $\sum_{i=-\infty}^{i=+\infty} \rho_{xx}(i)\rho_{yy}(i)/(N-k)$ where $\rho_{xx}(.), \rho_{yy}(.)$ denote the autocorrelation functions of the x- and y-series, respectively. (This is a special case of the general formula for computing standard errors of cross-correlations, usually called Bartlett's formula – see Box et al., 1994, equation (11.1.7).) The only case where the sample cross-correlation function is easy to interpret is when (i) the two processes are not cross-correlated and (ii) one process is white noise, and hence has no autocorrelations except at lag zero. In this case, the sample cross-correlation coefficient has an expected value which is approximately zero and a variance equal to $1/(N - k)$. Thus cross-correlations exceeding $\pm 2/\sqrt{(N - k)}$ may indeed be taken to be significantly different from zero. This test is used during the identification of transfer function models to assess the size of cross correlations between the input and output after the input has been transformed, or *prewhitened*, to white noise – see Section 5.2.2. However, it should be emphasized that testing cross-correlations based on the critical values $\pm 2/\sqrt{(N - k)}$ is not valid when both series exhibit non-zero autocorrelation, as is usually the case for the original observed data.

In view of the difficulties in interpreting the sample cross-correlation function, it usually needs to be supplemented by additional graphical and numerical tools. Here we present some additional graphical tools. As well as plotting the series one at a time, it can be fruitful to scale all the series to have the same variance, and then to plot pairs of variables on the *same* graph in order to reveal possible relationships between the series. Note that if the cross-correlation between two series at zero lag is negative, then one series should be turned 'upside-down' in order to get a good match, by reversing the sign of any difference from the overall mean. If one series lags the other, in that the behaviour of one series (such as the presence of a peak) follows similar behaviour in the other series after a constant time interval, then it may help to move one series forwards or backwards in time by an appropriate time interval in order to 'see' the relationship more clearly by making the visible characteristics of the series agree as closely as possible. Some trial-and-error may be needed to get an appropriate choice of scalings and of any time lags. A good interactive graphics package is essential to do all this. This strategy may be helpful in alerting the analyst to the possible presence of linear relationships, perhaps with a built-in delay mechanism. However, the approach suffers from the same drawback as attempts to interpret the sample cross-correlation function in that apparent relationships may be spurious because they are induced by autocorrelations within the series. Thus it is advisable to remove large autocorrelations from series (by applying the same transformation to both series) before looking at such time plots.

With two time series, an alternative possibility is to plot pairs of contemporaneous[1] observations on a single scatter plot, but then the resulting graph will give no indication of time ordering or of relationships at non-zero lags, unless each point on the graph is labelled in some way to indicate time or the graph is redrawn at selected lag differences. With more than two variables, modern graphics makes it easy to plot what is called a *scatterplot matrix*, which comprises a scatterplot of every pair of variables. This can also be difficult to interpret for time-series data.

---

[1] Contemporaneous measurements are those measured at the same time.

5.1.5 Some difficulties

The remarks in Section 5.1.4 have already indicated that there can be difficulties in analysing and interpreting multivariate time series. In my experience, the modelling process is fraught with danger, and this will become further evident in the rest of this chapter. It is difficult enough trying to fit a model to multivariate data which are *not* time series, and the complications induced by having time-ordered data naturally exacerbates these problems.

With a large number of variables, it is particularly difficult to build a good multivariate time-series model when some cross-correlations between the predictor variables are 'large'. Then it may be fruitful to make some sort of multivariate transformation of the data (e.g. by using principal component analysis or factor analysis) so as to reduce the effective dimensionality of the data. This type of approach will not be considered here – see, for example, Pena and Box (1987).

Apart from the difficulties in interpreting sample cross-correlations and related statistics, there are, more generally, many practical difficulties involved in *collecting* multivariate time-series data. Put crudely, the data that have been recorded may not be in the form that we would have preferred but we, nevertheless, have to make the best of them. There may of course be outliers and missing observations, which can be more prevalent, and even more difficult to deal with, than in the univariate case. Moreover, we might have liked data on some variables which have not been recorded or we might have preferred more controlled perturbations to input variables so as to better assess their effect on response variables. We might have preferred data recorded at different time intervals or to different levels of accuracy. All such issues should ideally have been considered prior to collecting the data, but, in practice, it is not always possible to prejudge such matters until some data have been collected, and then careful judgement may be needed to assess what can be done with the data.

5.2 Single-equation models

5.2.1 Regression models

A seemingly obvious class of single-equation models to apply to multivariate time-series data is the *multiple regression* model. This class of models is probably the most widely used in practice, especially in business, economics and the forecasting of public utilities, such as electricity. Regression models feature prominently in many texts on forecasting for management science and business students (e.g. Farnum and Stanton, 1989; Levenbach and Cleary, 1984; Makridakis and Wheelwright, 1989), partly because they can be used for data that are not time series. However, as outlined below, there can be serious problems in applying regression models to time-series data. Thus their use is often inadvisable or inappropriate, and they feature much less prominently in time-series books written by statisticians. Nevertheless, it seems advisable to include a full discussion of them, not only because they are widely used, but also because it is important to understand their drawbacks and because they provide a helpful lead-in to some alternative classes of model.

Revision of linear regression. We assume the reader has some familiarity with the linear regression for bivariate data, but include a brief summary for completeness. The standard model may be written

$$Y = \alpha + \beta x + \varepsilon$$

where Y denotes a response variable, x an explanatory variable, α, β are constant parameters, and the observation errors, $\{\varepsilon\}$, are usually assumed to be independent and normally distributed, with mean zero and constant variance σ^2. Alternatively we can write

$$E(Y|x) = \alpha + \beta x$$

$$\text{Var}(Y|x) = \sigma^2$$

so that observations on Y, for a given value of x, are independent $N(\alpha + \beta x, \sigma^2)$. The values of the explanatory variable, x, are often assumed to be, in principle at least, under the control of the experimenter and explains why the variable is written here in lower case (in contrast to the response variable, Y, which is a random variable and so is written in capitals). Given n pairs of observations, say $\{(y_i, x_i) \text{ for } i = 1, 2 \ldots n\}$, least squares estimates of α and β may be found as

$$\hat{\alpha} = \bar{y} - \hat{\beta}\bar{x}$$

$$\hat{\beta} = \sum x_i(y_i - \bar{y})/ \sum (x_i - \bar{x})^2$$

We also assume some familiarity with the generalization of the above to the case where there are p explanatory variables, say $\boldsymbol{x}^T = (x_1, x_2, \ldots, x_p)$. If the conditional expectation of Y is linearly related to $\boldsymbol{x}$, then we write

$$E(Y|\boldsymbol{x}) = \beta_1 x_1 + \beta_2 x_2 + \ldots + \beta_p x_p = \boldsymbol{\beta}^T \boldsymbol{x} \qquad (5.2.1)$$

where $\boldsymbol{\beta}^T = (\beta_1, \ldots, \beta_p)$ is the vector of regression coefficients. A constant (intercept) term can be included by setting one of the x-variables to be a constant. The deviations from the conditional mean are also generally assumed to be independent $N(0, \sigma^2)$. Given n sets of observations, say $\{(y_i, x_{i1}, \ldots, x_{ip}) \text{ for } i = 1, 2, \ldots, n\}$, least squares estimates of $\boldsymbol{\beta}$ are given by

$$\hat{\boldsymbol{\beta}} = (X^T X)^{-1} X^T \boldsymbol{y} \qquad (5.2.2)$$

where the $(n \times p)$ matrix X, often called the *design matrix*, has (i, j)th element x_{ij} and $\boldsymbol{y}^T = (y_1, \ldots, y_n)$. Further details may be found in many statistics textbooks.

Application to time-series data. Considerable care is needed to apply regression models to time-series data. If, for example, we simply write a regression model relating a response variable, Y_t, to a single explanatory variable, x_t, in the form

$$Y_t = \alpha + \beta x_t + \varepsilon_t$$

with successive values of ε_t assumed to be independent, then this model fails to capture the temporal dependence among the observed variables and among the 'error' terms. In other words, it ignores the fact that the variables are all ordered through time. With time-series data, the *order* of the data is usually crucial, because successive values are typically (auto)correlated and this needs to be taken into account when modelling the data. Failure to do so can produce a flawed model.

Some adjustment to notation is necessary to cope with time-series data We use t, rather than i, as the dummy subscript and N, rather than n, for the length of the series. With p explanatory variables, say $x_1, x_2, \ldots, x_p$, it is customary in the time-series literature to reverse the order of the subscripts of the x-variables so that the data are written as $\{(y_t, x_{1t}, x_{2t}, \ldots, x_{pt})$ for $t = 1, \ldots, N\}$. In other words, the first subscript now denotes the relevant variable. Given that most models will involve lagged values, it is generally impossible to make the subscripts correspond to the position in the design matrix as in ordinary multiple regression, and so there is no advantage in adopting the conventional regression notation.[2]

When modelling time-series data, it is helpful to distinguish between the various types of measured explanatory variables which can arise. One distinction is between variables measured at the same time as the response variable, called contemporaneous variables, and those measured at an earlier time, called lagged variables. The latter may include past values of the explanatory variables and of the response variable. In this case the number of usable multivariate data points will be less than N. Lagged values of explanatory variables can be regarded as leading indicators – see (5.1.1) – and they can help to improve the forecasting ability of the model. However, the inclusion of 'explanatory variables' which are actually lagged values of the response variable (i.e. autoregressive terms) will change the character of the model substantially, as such variables are almost certainly correlated with each other and with the response variable, and will often explain much of the variation in the response variable before any explanatory variables are introduced.

Another distinction is between variables which can, or which cannot, be controlled. This is particularly important when trying to model data where feedback is present. A further distinction is between explanatory variables which are thought to have some *causal* effect on the response variable and those that do not. The former are clearly more reliable for making forecasts, as, for example, when orders are used to forecast production. Indeed Clements and Hendry (1998a, Chapter 9) suggest that it is unwise to use explanatory variables unless there is a causal relationship, as they say that "leading indicator systems are altered sufficiently frequently to suggest that they do not systematically lead for long". A final distinction is between measured explanatory variables and those that are predetermined quantities such as a deterministic function of time itself. For example, a

---

[2] Of course, the ordering of the subscripts is arbitrary as long as the treatment is self-consistent. The notation adopted here is fairly standard in the time-series literature and has some advantages.

linear regression on time can be obtained by setting $x_{1t} = kt$, where k is a constant. Note that regression on time alone is normally regarded as a univariate procedure.

The time-series version of the multiple regression model (c.f. (5.2.1)) may be written

$$Y_t = \beta^T x_t + u_t \qquad\qquad (5.2.3)$$

where x_t denotes the vector of explanatory variables at time t and u_t denotes the observation error at time t. At first sight, it may look straightforward to fit this model to time-series data. If we can assume that x_t are predetermined variables which do not include any lagged variables (either of the response or of the explanatory variables), and if we can also assume that the u_t are independent $N(0, \sigma^2)$ variables which are uncorrelated with the values of x_t, then it is indeed easy to fit the model by least squares using (5.2.2). This is called fitting by *Ordinary Least Squares*, which is abbreviated OLS in the econometric literature. However, as noted above, these assumptions are highly unlikely to apply to time-series data. In particular, the observation errors are likely to be correlated and this xplains why we use the notation u_t, rather than ε_t.

Econometricians have developed some alternative estimation procedures to cope with various different assumptions that could reasonably be made. For example, *Generalized Least Squares* (GLS) may be used when the u_t are assumed to be (auto)correlated with a known structure, while an iterative procedure called the *Cochrane-Orcutt procedure* may be used when the u_t are thought to follow an AR(1) process but with an unknown parameter. If the u_t are thought to be correlated with some of the x_t variables, then a procedure called *two-stage least squares* may be appropriate. Some theoretical results are also available to handle further departures from the basic assumptions of ordinary multiple regression, such as the case where the explanatory variables are stochastic, rather than predetermined, and where the innovations have a non-Gaussian distribution. A overview of the many approaches to time-series regression is given by Choudhury et al. (1999), and the reader is referred to an econometric text such as Hamilton (1994) for more details.

Having said all this, it is my experience that forecasters rarely use such modifications, but rather use OLS, rightly or wrongly, even when the necessary assumptions do not apply, partly because the necessary software is readily available and easy to use, and partly because it may not be easy to decide what alternative should be used. It is therefore worth considering the problems that may arise when fitting regression models to time-series data, and see why such models are viewed with scepticism by most statisticians with experience in time-series analysis.

More problems. Given the wide availability of suitable software, it is computationally easy to fit any given regression model to any given set of data. However, building a successful model, when the structure is unknown a priori (as it usually will be), is anything but easy, especially for time-series data. An iterative procedure is usually adopted, trying (i) different subsets of the available explanatory variables, (ii) different lagged structures, (iii) different transformations of the variables, (iv) fits with and without any outliers; and so on. After doing all this, it is comparatively easy to get a good fit to virtually any given set of data. For example, Armstrong (1970) describes an alarming study where a plausible multiple regression model was fitted to data which were actually completely random. Armstrong (1985, p. 487) later describes some rules for 'cheating' to get a high (within-sample) multiple correlation coefficient, R^2, by (i) omitting 'annoying' outliers which degrade the fit, (ii) using R^2 as the measure of fit rather than a statistic, like the adjusted-R^2-statistic or AIC, which takes account of the number of variables fitted, (iii) including lots of variables. As regards (iii), in business and commerce, it is easy to think of many plausible explanatory variables to include and the ready availability of easy-to-use subset selection software, has led to excessive numbers of explanatory variables being tried in some modelling exercises. With up to, say, 20 explanatory variables it is often possible to get a value of R^2 close to one. However, a good (within-sample) fit does not necessarily translate into good out-of-sample forecasts and regression models fitted to time-series data, especially when non-stationary, may be spurious, as can readily be demonstrated both empirically (Granger and Newbold, 1974; 1986, Section 6.4) and theoretically (Phillips, 1986; Hamilton, 1994, Section 18.3).

The difficulties in applying multiple regression to time-series data often relate to the quality and structure of the data. These difficulties can be particularly acute with economic data. If, for example, the explanatory variables are highly correlated with each other, as well as autocorrelated through time, there can be singularity problems when inverting the $(X^T X)$ matrix. In such situations it is misleading to refer to explanatory variables as being the *independent* variables, and this terminology, widely used before about 1980, has still not completely lapsed, especially in the economic literature. It is advisable to look at the correlation matrix of the explanatory variables before carrying out a multiple regression so that, if necessary, some variables can be transformed or excluded. It is not necessary for the explanatory variables to be completely independent, but large correlations should be avoided.

Another difficulty with some observed regression data is that a crucial explanatory variable may have been held more or less constant in the past so that it is impossible to assess its quantitative effect accurately.[3] For example, a company may be considering increasing its advertising budget and would like a model to predict the likely effect on sales. Unfortunately, if advertising has been held more or less constant in the past, then it is not possible to estimate empirically the effect on sales of a change in advertising. Yet a model which excludes the effect of advertising could be useless if the advertising budget were to be changed. Thus it may be

---

[3] This point was briefly introduced in Section 1.2.

necessary to appeal to economic or business theory during the modelling process.

Yet another problem in modelling multivariate data is that there may be feedback from the response to the explanatory variables (see Section 5.1.1), in which case it is unwise to try to fit any single-equation model, but rather it will be necessary to set up a model with more than one equation such as the vector autoregressive model (see Section 5.3) or an econometric simultaneous equation model (see Section 5.5).

The problems listed above may generally be described as resulting from the unsuitable nature of some datasets for fitting regression models. However, the final problem noted here relates to the basic unsuitability of the model, as mentioned earlier in this section. The usual regression model fails to take account of the temporal dependence in time-series data. In particular, a key reason why spurious multiple regression models may be constructed is that the assumed error structure may be overly simplistic for describing time-series data. The usual multiple regression model assumes that observation errors are independent and hence uncorrelated through time. Unfortunately, this is usually not the case for time-series data.

A celebrated example is discussed by Box and Newbold (1971). An earlier analysis of some financial data appeared to show a relationship between the *Financial Times* ordinary share index and the U.K. car production six months earlier. This resulted in a leading indicator-type model of the form given by (5.1.1) with observation errors that were assumed to be uncorrelated. The identification of this model relied heavily on the 'large' cross-correlations which were observed between the two series at a lag of six months. In fact, the latter were a direct result of the autocorrelations *within* the series, rather than from any relationship *between* the series, and Box and Newbold (1971) showed that the two series involved were essentially uncorrelated random walks. In the earlier analysis, the failure to difference the series to make them stationary, before calculating the cross-correlations, was unwise and meant that the variance of the cross-correlations was highly inflated. This illustrates the difficult in interpreting cross-correlations – see also Section 5.1.4. Box and Newbold (1971) went on to stress the importance of making diagnostic checks on any fitted model, including an analysis of the (one-step-ahead forecast) residuals. If the latter are found to be autocorrelated, then an alternative model should be sought.

Some precautions. There are various precautionary measures which can be taken to try to avoid spurious regressions. They include the following:

1. Choose the explanatory variables with great care using external knowledge. Limit the maximum number of variables to perhaps five or six.

2. Consider including lagged values of both the response and explanatory variables as necessary.

3. Remove trend and seasonal variation before carrying out regression.

Consider the use of differencing to make data stationary, but note that this can lead to misspecification if a phenomenon called *cointegration* is present – see Section 5.4.

4. Carry out careful diagnostic checks on any fitted model. In addition, it is generally advisable to divide the data into two parts, fit the model to the first (and usually larger) part of the data and then check it by making forecasts of the later part. Wherever possible, such forecasts should be genuine out-of-sample forecasts – see Section 5.1.2.

5. Allow for correlated errors by using a more sophisticated fitting procedure, such as Generalized Least Squares – see above.

However, even with the above precautions, it can still be dangerous to use multiple regression models for time-series forecasting. Thus the best precaution may be to avoid using regression models at all! There are various alternative classes of models, which will generally be preferred to regression models. Transfer function models – see Section 5.2.2 – are usually a safer option for a single-equation model, as they allow a more general error structure and a more general lead-lag relationship. Alternatively, if a model comprising more than one equation is indicated (e.g. if feedback is thought to be present in the data), then there are various classes of models available – see Sections 5.3 to 5.5.

In summary, the use of multiple regression models, with uncorrelated errors, for describing time-series data is generally not recommended.

5.2.2 Transfer function models

This class of single-equation models generally provides a sounder basis for modelling time-series data than regression models. The general structure of such models may be introduced by noting that the most useful regression model for forecasting time series appears to be the leading indicator type of model given in (5.1.1), namely

$$Y_t = a + bX_{t-d} + \varepsilon_t \tag{5.2.4}$$

where a, b, d are constants, with d a positive integer. Note that the explanatory variables will henceforth be written in capitals, rather than lower case, as they may be stochastic, rather than predetermined. This model enables forecasts of Y_t to be made for up to d steps ahead without having to forecast the explanatory variable.

However, some of the problems listed in Section 5.2.1 will sadly apply to models of this type. In particular, the observation errors, the $\{\varepsilon_t\}$, are unlikely to be independent, as typically assumed in regression, but rather are likely to be autocorrelated in a way which is unknown a priori. Moreover, if X_t has a (delayed) effect on the response variable, there is no obvious reason why it should only be at one particular time lag (even if we can assume that there is no feedback from the response variable to the explanatory variable). Thus it is usually safer to try to fit an alternative class of models, which allows the observation errors to be autocorrelated and which also allows the explanatory variable to affect the response variable in more than one time period. This suggests trying the following equation, namely

$$Y_t = \nu_0 X_t + \nu_1 X_{t-1} + \ldots + n_t = \nu(B)X_t + n_t \qquad (5.2.5)$$

where $\nu(B) = \nu_0 + \nu_1 B + \nu_2 B^2 + \ldots$ is a polynomial in the backward shift operator, B, and $\{n_t\}$ denotes the observation errors which are sometimes called the 'noise' in this context. The latter series is generally *not* an independent series, but it is necessary to assume that it is uncorrelated with the input series. This model is called a *transfer function-noise model* or, more simply, a *transfer function model*. Such models provide a safer, more general class of models than regression models, to enable the forecaster to describe the dynamics of an open-loop causal relationship between a single predictor variable and a single response. The input has an effect on the output over a possibly infinite number of lags. As a result, models of this type are often called (free-form) *distributed lag models* by econometricians, because they 'distribute the lead-lag relationship' over many lags.

Two further points can be made about (5.2.5). Firstly, notice that there is no constant on the right-hand side of the equation as there is in (5.2.4), for example. It is possible to add a constant but it is not customary to do so, partly because it complicates the mathematics, and partly because the data are usually analysed in a form which makes it unnecessary, as for example if the variables are mean-corrected, or are deviations from equilibrium. Secondly, note that if ν_0 is not zero, then there is a contemporaneous relationship between X_t and Y_t and, in order to avoid feedback, the analyst must be prepared to attribute this relationship to a dependence of Y on X, rather than vice versa.

Readers with a general knowledge of linear systems will recognize (5.2.5) as being a (physically realizable or causal) linear system plus noise. (An introduction to linear systems is given for example by Chatfield, 1996a, Chapter 9.) Looking at the linear system part of the model, namely

$$Y_t = \nu(B)X_t \qquad (5.2.6)$$

it turns out that the sequence of weights, namely $\nu_0, \nu_1, \nu_2, \ldots$, constitute what is called the *impulse response function* of the system, because ν_t is the response of the system at time t to an input series which is zero except for an impulse of unity at time zero. If the first few values of the impulse response function are zero, then X_t is a leading indicator for Y_t. In particular, if $\nu_0 = \nu_1 = \ldots = \nu_{d-1} = 0$, then there is a delay, d, in the system and $\nu(B) = \nu_d B^d + \nu_{d+1} B^{d+1} + \ldots = B^d(\nu_d + \nu_{d+1}B + \ldots)$. Then it may be convenient to write (5.2.6) in the form

$$Y_t = \nu^*(B)X_{t-d} \qquad (5.2.7)$$

or equivalently as

$$Y_t = \nu^*(B)B^d X_t \qquad (5.2.8)$$

where $\nu^*(B) = \nu_d + \nu_{d+1}B + \ldots$.

Note that Box et al. (1994) restrict the use of the phrase *transfer function model* to models as in Equation (5.2.6) to distinguish them from transfer function-noise models as in (5.2.5), which have noise added, but most other writers use the term *transfer function* model for models with or without noise. We generally adopt the latter usage, while making clear when noise is present or not.

The transfer function model is typically applied to data in the physical sciences, where feedback is unlikely to be a problem. It is also worth noting that the general linear process in Equation (3.1.14) is an MA model of infinite order and hence can be thought of as a linear system, or as a transfer function model, which converts an 'input' consisting of white noise, $\{Z_t\}$, to an 'output', namely the measured series, $\{X_t\}$.

We now explore the possibility of writing a transfer function model in a rather different way. Recall that a general linear process can often be parsimoniously represented as a mixed ARMA process of low order. This means that, instead of working with a model which has a possibly infinite number of parameters, the analyst can use a model where the number of parameters is finite and usually quite small. By the same token, (5.2.5) has an infinite number of parameters which will be impossible to estimate from a finite sample. Moreover, even if the parameter values were available, it would be impossible to evaluate terms in the linear sum involving observations before the beginning of the observed data. Fortunately, there are mathematical theorems which tell us that an infinite-order power series, like $\nu(x)$ (where we replace the operator B with a real variable x), can usually be approximated to an arbitrary degree of accuracy by the *ratio* of two finite-order polynomials, say $\omega(x)/\delta(x)$, provided that the ν_i coefficients converge in an appropriate way.[4]

Writing $\delta(B) = (1 - \delta_1 B - \ldots - \delta_r B^r)$ and $\omega(B) = (\omega_0 + \omega_1 B + \ldots + \omega_s B^s)$, the above remarks suggest that the transfer function model in (5.2.7) or (5.2.8) can usually be parsimoniously rewritten in the form

$$\delta(B)Y_t = \omega(B)X_{t-d} = \omega(B)B^d X_t \qquad (5.2.9)$$

where $\delta(B)$, $\omega(B)$ are low-order polynomials in B, of order r, s, respectively, such that $\omega(B) = \delta(B)\nu^*(B)$. This follows the notation of Box et al. (1994) except that, like most other authors, the coefficients in $\omega(B)$ are taken to be positive so that the sign of ω_i corresponds to the sign of the change in Y for a positive change in X. In this revised form, (5.2.9) may include lagged values of Y_t as well as of X_t. The same is true of the corresponding transfer function-noise model. As the ratio of two polynomials is called a *rational function* by mathematicians, (5.2.9) is often called a *rational distributed lag model* by econometricians.

---

[4] Note that a sufficient condition for the linear system to be *stable* is that $\sum |\nu_i|$ is finite, but we will not go into the mathematical details here.

Example 5.1. This example demonstrates that (5.2.9) may provide a much simpler working formula than the general linear system in (5.2.6) or (5.2.7). Suppose that a change in the input has no effect for the first three time periods, but that, thereafter, there is an exponentially decaying effect, such that $\nu_3 = \lambda$, $\nu_4 = \lambda^2$ and so on. Then $\nu^*(x) = \lambda \sum_{i=0}^{\infty}(\lambda x)^i = \lambda/(1 - \lambda x)$ so that $\nu^*(x)$ is the ratio of a constant and a first-order polynomial. Then, taking $\delta(B) = (1 - \lambda B)$ and $\omega(B) = \lambda$, we may write (5.2.9) in the form

$$(1 - \lambda B)Y_t = \lambda X_{t-3} \qquad (5.2.10)$$

or

$$Y_t = \lambda(Y_{t-1} + X_{t-3}) \qquad (5.2.11)$$

This is a nice simple form with just one parameter to estimate once the values $d = 3$, $r = 1$ and $s = 0$ have been identified. More generally, if the $\{\nu_i\}$ decay exponentially from the general value, ν_3, so that $\nu_i = \nu_3 \lambda^{i-3}$ for $i \geq 3$ (and zero otherwise), then (5.2.11) becomes

$$Y_t = \lambda Y_{t-1} + \nu_3 X_{t-3} \qquad (5.2.12)$$

This is a special case of the *Koyck* econometric model. □

If we multiply through (5.2.9) by $\delta^{-1}(B)$ and then add noise as in (5.2.5), we get

$$Y_t = \delta^{-1}(B)\omega(B)X_{t-d} + n_t \qquad (5.2.13)$$

If we can model the noise as an ARIMA process, say

$$\phi_n(B)n_t = \theta_n(B)Z_t \qquad (5.2.14)$$

where Z_t denotes a white noise process, then we may further rewrite (5.2.13) in the form

$$Y_t = \delta^{-1}(B)\omega(B)X_{t-d} + \phi_n^{-1}(B)\theta_n(B)Z_t \qquad (5.2.15)$$

If the data have been differenced before the start of the analysis, then an ARMA, rather than ARIMA, model will probably be adequate for modelling the noise. Otherwise, $\phi_n(B)$ may have roots on the unit circle.

Note that (5.2.15) reduces to a simple linear regression model, with a leading indicator as input, when $\delta(B) = \phi_n(B) = \theta_n(B) = 1$ and $\omega(B) = \nu_d B^d$. Note also that if n_t in (5.2.13) is white noise, then the model can be regarded as a regression model with both lagged dependent and explanatory variables. As a final special case of the transfer function model, consider (5.2.5), where $\nu(B)$ is a polynomial of finite order and n_t follows an AR(1) process. Then the model can be regarded as regression with (first-order) autoregressive 'errors' for which the Cochrane-Orcutt estimation procedure can be used (see Section 5.2.1).

Fitting a transfer function model. Given this book's focus on forecasting, only a brief summary of the (rather complicated) procedure for fitting transfer function models will be given here. Full details may be found in Box et al. (1994, Chapters 10 and 11). Note that there are several minor variants to the procedure, although these alternative procedures are generally closely related to the Box-Jenkins approach. For example, Pankratz (1991) describes a procedure based on a class of models called *dynamic regression models* which appear to be equivalent to transfer function-noise models.

It is important to begin the analysis, as always, by looking at time plots of the data to check for any errors or outliers, so that, if necessary, the data can be 'cleaned'. The next step in the analysis is usually to transform the input and output series, if necessary, in order to make them stationary in the mean (and more rarely in the variance). This is usually accomplished by differencing (and applying a power transformation), but note that there could be occasions when it is unwise to difference both series as for example if external theory suggests that the *level* of Y is related to the *change* in X. Both series can also be mean-corrected, and, in the remainder of this section, X_t and Y_t denote the transformed series which are assumed to have mean zero.

At first sight, the next step would be to calculate the cross-correlation function of the input and output. However, we have seen that this function is very difficult to interpret while autocorrelations exist within both series, primarily because the variance of cross-correlations is inflated by autocorrelations. The usual way of overcoming this problem is to fit an ARIMA model to the input, so that

$$\phi_x(B)X_t = \theta_x(B)\alpha_t \qquad (5.2.16)$$

where α_t denotes a white noise process. This can be regarded as a way of transforming the input series, X_t, to white noise, a procedure that is sometimes called *prewhitening*. The same transformation that has been applied to do this, namely $\phi_x(B)\theta_x^{-1}(B)$, is now applied to the output series, Y_t, to give

$$\phi_x(B)\theta_x^{-1}(B)Y_t = \beta_t \qquad (5.2.17)$$

This means that (5.2.5) may be rewritten in the form

$$\beta_t = \nu(B)\alpha_t + \eta_t \qquad (5.2.18)$$

where η_t is the transformed noise series, namely $\phi_x(B)\theta_x^{-1}(B)n_t$. The big advantage of expressing the model as in (5.2.18) is that the cross-correlation function of the two transformed series is now much easier to interpret, as it can readily be shown (e.g. Box et al, 1994, Section 11.2.1) that it is now directly proportional to the impulse response function. Thus, if $r_{\alpha\beta}(k)$ denotes the sample cross-correlation coefficient between the transformed series $\{\alpha_t\}$ and $\{\beta_t\}$ at lag k, then the natural estimate of ν_i turns out to be given by

$$\hat{\nu}_i = r_{\alpha\beta}(i)s_\beta/s_\alpha \qquad (5.2.19)$$

where s_α, s_β denote the observed standard deviations of the transformed series. The pattern of the estimated ν-weights is compared with some common theoretical patterns of ν-weights for different low-order rational models of the form in (5.2.9) and this should enable the analyst to identify appropriate values of the integers r, s and d, and hence fit an appropriate model of the form in (5.2.9). As always, it is important to use any external knowledge in doing this, as, for example, if there is a known delay time. If the latter is not known, then the first few values of $\hat{\nu}_i$ should be tested to see if they are significantly different from zero by comparing the value of $r_{\alpha\beta}(i)$ with $\pm 2/\sqrt{(N-i)}$ – see Section 5.1.4. A simple way to identify the value of d, the delay, is to choose the smallest integer, k, such that $|r_{\alpha\beta}(k)| > 2/\sqrt{(N-k)}$. Having identified preliminary values of d, r and s,

preliminary estimates of the parameters of the model in (5.2.9) can be found
using (5.2.19).

The analyst also needs to identify an appropriate ARIMA model for the
noise as in (5.2.14). A first assessment is made by computing the residuals
from the preliminary estimate of the model in (5.2.13), and identifying
an appropriate ARIMA model for these preliminary residuals by the usual
(univariate) identification procedure. Then the whole model of the form in
(5.2.15) can be refitted, estimating the parameters of (5.2.9) and of the noise
model in (5.2.14), all at the same time. The residuals from this re-estimated
model of the form in (5.2.13) can now be (re)calculated and used to check that
the structure of the ARIMA model for the noise still seems appropriate. If checks
indicate that a different ARIMA model should be tried, then the whole model
will need to be refitted once again. This iterative process continues until the
analyst is satisfied with the results of appropriate diagnostic checks which should
include checking the residuals from the ARIMA noise model, and also checking
the cross-correlations between (i) the residuals from the ARIMA noise process
and (ii) the residuals, α_t, which arise when fitting an ARIMA model to the input
as in (5.2.13). In the absence of feedback, these two series should be uncorrelated.

Forecasting with a transfer function model. When the analyst is eventually
satisfied with the fitted model, it may be used in forecasting. The
computation of forecasts for (5.2.13) is carried out in an 'obvious' way
by computing forecasts for the linear system part of the model (5.2.9),
and also for the noise part of the model (5.2.14), and then adding them
together. The rules for doing this, so as to compute the minimum mean
square error (MMSE) prediction, are natural multivariate extension of the
rules for computing MMSE forecasts for an ARIMA model – see Section
4.2.1. Forecast formulae are found by writing out the model equation and
substituting observed values as follows:

(i) Present and past values of the input and output variables are replaced
by their observed values.

(ii) Future values of the output, Y_t, are replaced by their conditional
expectation or MMSE forecast from the full transfer-function noise
model.

(iii) Future values of the input, X_t, are replaced by their conditional
expectation or MMSE forecast using the univariate model fitted to the
input as in (5.2.16).

(iv) Present and past values of n_t are obtained from (5.2.13) as the residual
from the linear system part of the forecast, namely

$$n_t = Y_t - \delta^{-1}(B)\omega(B)X_{t-d}$$

(v) Future values of n_t are replaced by their MMSE prediction using the
ARIMA model in (5.2.14).

(vi) If needed, present and past values of Z_t can be found from the noise
series, n_t, using (5.2.14) via $Z_t = \theta_n^{-1}(B)\phi_n(B)n_t$, or more simply as

$Z_t = n_t - \hat{n}_{t-1}(1)$ or, equivalently, as the overall one-step-ahead forecast error using the complete equation, (5.2.15), namely

$$Z_t = Y_t - \hat{Y}_{t-1}(1)$$

where $\hat{Y}_{t-1}(1)$ is the overall forecast including the forecast of the linear system part of the model and the forecast of the noise component. Of course, future values of Z_t are best predicted as zero.

(vii) In practice, the model will not be known exactly so that model parameters have to be estimated, giving estimates of n_t, Z_t, and of the conditional expectations that can be substituted into the appropriate formulae.

Example 5.1 continued. The above procedure for computing forecasts will be illustrated using the Koyck model in (5.2.12), together with added noise which is assumed to follow an AR(1) process with parameter ϕ_n. Then the model expressed in the form of (5.2.15) is

$$Y_t = (1 - \lambda)^{-1} \nu_3 X_{t-3} + n_t$$

where $n_t = (1 - \phi_n B)^{-1} Z_t$. We look separately at forecasts of the two components of this model.

At time N, the MMSE h-step-ahead forecast of the AR(1) noise process, n_t, is given by

$$\hat{n}_N(h) = \phi_n^h n_N$$

In this case, it is unnecessary to find the values of Z_t that correspond to the $\{n_t\}$ series in order to compute predictions, though it is probably advisable to look at them in order to check that the structure of the model is adequate. However, if n_t followed an MA or ARMA process, rather than an AR process, then the values of Z_t would be needed for prediction as well as model-checking.

When computing forecasts for the transfer function part of the model, we rewrite (5.2.9) in the form of (5.2.12). Thus, if there were no noise, the one-step-ahead forecast at time N would be given by

$$\hat{Y}_N(1) = \lambda Y_N + \nu_3 X_{t-2} \qquad (5.2.20)$$

where the values on the right-hand side of the equation are all known. However, to forecast two steps ahead using (5.2.12), we find

$$\hat{Y}_N(2) = \lambda Y_{N+1} + \nu_3 X_{t-1} \qquad (5.2.21)$$

where the value of Y_{N+1} is of course not available at time N. Then, using the general guidelines, future values of the output should be replaced by MMSE forecasts (as is intuitively 'obvious'?) and the forecast of Y_{N+1} has already been computed in (5.2.20). Putting this into (5.2.21), gives

$$\hat{Y}_N(2) = \lambda \hat{Y}_N(1) + \nu_3 X_{t-1} \qquad (5.2.22)$$

A forecast for Y_{N+3} can be similarly computed using $\hat{Y}_N(2)$ and this recursive way of calculating forecasts is common practice. To forecast four steps ahead, the value of X_{N+1} is needed but is also not available. Using the general guidelines again, this value should be replaced with the ARMA forecast based on the model fitted to the input (see (5.2.16)) as part of the model-identification process. Suppose, for example, that the input follows an AR(1) process with parameter ϕ_x. Then the best forecast of X_{N+1} made at time N is given by $\hat{X}_N(1) = \phi_x X_N$. Forecasts further ahead can be computed by $\hat{X}_N(h) = \phi_x^h X_N$.

Having found how to compute forecasts for the linear system part of the model, we can now combine them with forecasts of the noise. For the Koyck model with AR(1) noise and AR(1) input, the forecasts are calculated recursively and we find, for example, that the MMSE forecasts for one and four steps ahead are

$$\hat{Y}_N(1) = \lambda Y_N + \nu_3 X_{t-2} + \phi_n n_N$$

$$\hat{Y}_N(4) = \lambda \hat{Y}_N(3) + \nu_3 \hat{X}_N(1) + \hat{n}_N(4) = \lambda \hat{Y}_N(3) + \nu_3 \phi_x X_N + \phi_n^4 n_N$$

If the original data were differenced or otherwise transformed, forecasts of the original observed variable can readily be found from the above.

In practice, the model parameters, and the values of n_t will not be known exactly and so must be replaced with estimates. The one-step-ahead forecast, for example, will be $\hat{Y}_N(1) = \hat{\lambda} Y_N + \hat{\nu}_3 X_{t-2} + \hat{\phi}_N \hat{n}_N$ □

Note that it can also be helpful to fit a univariate ARIMA model to the original output (as well as fitting an ARIMA model to the input). This univariate model can be used for comparison purposes. If the transfer function model cannot give a better fit to the output, or more importantly better out-of-sample forecasts than the univariate model, then it is probably not worth using. However, there are now many convincing examples in the literature (e.g. Jenkins, 1979) to demonstrate that transfer function models can lead to improved performance.

5.3 Vector AR and ARMA models

When data are generated by a closed-loop system (see Section 5.1.1), it no longer makes much sense to talk about an 'input' and an 'output', and a single-equation model will no longer be adequate to describe the data. Subsequent sections consider the case where there are two, or more, variables which are all interrelated and which are sometimes described as 'arising on an equal footing'. Modelling a set of interrelated variables is often called *multiple time-series modelling*. Econometricians would say that instead of having one endogenous variable (the output) and one or more exogenous variables (the input), we have two or more endogenous variables. Roughly speaking, an *exogenous* variable is one which affects the system but is not affected by it, whereas an *endogenous* variable *is* affected by the system and is therefore typically correlated with the series of observation errors.

With m variables which arise on an equal footing, we use the notation of Section 5.1.3, namely that we observe an $(m \times 1)$ vector $\boldsymbol{X}_t$ where $\boldsymbol{X}_t^T = (X_{1t}, \ldots, X_{mt})$. Note that there is no particular reason to label one variable as the 'output', Y_t, as in Section 5.2. For simplicity we give examples concentrating on the bivariate case when $m = 2$ and $\boldsymbol{X}_t^T = (X_{1t}, X_{2t})$.

An 'obvious' way to define a useful class of linear multivariate processes is to generalize the class of univariate autoregressive-moving average (ARMA) models, introduced in Section 3.1, by requiring that each component of $\boldsymbol{X}_t$ should be a linear sum of present and past values of (all) the variables and of (all) the white noise innovations. The latter are the basic building block of many multivariate time-series models and we therefore begin by defining multivariate white noise.

5.3.1 Multivariate white noise

Let $\boldsymbol{Z}_t^T = (Z_{1t}, Z_{2t}, \ldots, Z_{mt})$ denote an $(m \times 1)$ vector of random variables. This multivariate time series will be called *multivariate white noise* if it is stationary with zero mean vector, $\boldsymbol{0}$, and if the values of $\boldsymbol{Z}_t$ at different times are uncorrelated. Thus the covariance matrix function (see Section 5.1.3) of $\boldsymbol{Z}_t$ is given by

$$\Gamma(k) = \left\{ \begin{array}{ll} \Gamma_0 & k = 0 \\ 0_m & k \neq 0 \end{array} \right.$$

where Γ_0 denotes an $(m \times m)$ symmetric positive-definite matrix and 0_m denotes an $(m \times m)$ matrix of zeroes.

Notice that all the components of $\boldsymbol{Z}_t$ are uncorrelated with all the components of $\boldsymbol{Z}_s$ for $t \neq s$. Thus each component of $\boldsymbol{Z}_t$ behaves like univariate white noise. Notice also that the covariance matrix at lag zero, namely Γ_0, need not be diagonal (though it often is). It could happen that a measurement error at a particular time point affects more than one measured variable at that time point. Thus it is possible for the components of $\boldsymbol{Z}_t$ to be contemporaneously correlated.

It is sometimes necessary to make a stronger assumption about the values of $\boldsymbol{Z}_t$ at different time points, namely that they be *independent*, rather than just uncorrelated. As in the univariate case, independence implies zero correlation, but the reverse need not be true (although it is when $\boldsymbol{Z}_t$ follows a multivariate normal distribution).

5.3.2 Vector ARMA models

A mathematically succinct way of representing a class of linear multivariate models is to generalize the univariate ARMA model by writing

$$\Phi(B)\boldsymbol{X}_t = \Theta(B)\boldsymbol{Z}_t \tag{5.3.1}$$

where $\boldsymbol{X}_t$ and $\boldsymbol{Z}_t$ are m-vectors, and $\Phi(B), \Theta(B)$ are matrix polynomials in the backward shift operator B of order p and q, respectively. Thus we take $\Phi(B) = I - \Phi_1 B - \ldots - \Phi_p B^p$ and $\Theta(B) = I + \Theta_1 B + \ldots + \Theta_q B^q$, where I is the $(m \times m)$ identity matrix and $\Phi_1, \Phi_2, \ldots \Phi_p$ and $\Theta_1, \Theta_2, \ldots, \Theta_q$ are $(m \times m)$ matrices of parameters. If $\boldsymbol{Z}_t$ denotes multivariate white noise, then the observed m-vector of random variables, $\boldsymbol{X}_t$, is said to follow a *vector ARMA* (or VARMA) model of order (p, q). Note that (5.3.1) reduces to the familiar univariate ARMA model when $m = 1$. We restrict attention to *stationary* multivariate processes, and hence, without loss of generality, we may assume the variables have been scaled to have zero mean. The condition for stationarity is that the roots of the equation: determinant$\{\Phi(x)\} = |I - \Phi_1 x - \Phi_2 x^2 - \ldots - \Phi_p x^p| = 0$, should lie outside the unit circle. Note that this condition reduces to the familiar condition for stationarity in the univariate case when $m = 1$.

There is a corresponding condition for *invertibility*, namely that the roots of the equation: determinant$\{\Theta(x)\} = |I + \Theta_1 x + \Theta_2 x^2 + \ldots + \Theta_q x^q| = 0$, should lie outside the unit circle. This condition also reduces to the corresponding univariate condition when $m = 1$.

One problem with VARMA models is that there may be different, but equivalent (or exchangeable), ways of writing what is effectively the same model (meaning that its correlation structure is unchanged). There are various ways of imposing constraints on the parameters involved in (5.3.1) to ensure that a model has the property of being *identifiable*, meaning that we can uniquely determine the order of the model (i.e. the values of p and q) and the matrices of coefficients (i.e. $\Phi_1, \ldots, \Phi_p$ and $\Theta_1, \ldots, \Theta_q$), given the correlation matrix function of the process. However, the conditions required for uniqueness of representation are complicated and will not be given here.

The definition of a VARMA model involves matrix polynomials, which may look rather strange at first. However, they can readily be clarified with examples. For simplicity, we begin by considering *vector autoregressive processes*.

5.3.3 VAR models

Vector autoregressive models (usually abbreviated as VAR models) arise when there are no moving average terms in (5.3.1) so that $\Theta(B) = I$ and $q = 0$. As in the univariate case, many people find AR processes intuitively more straightforward to understand than MA processes, and, in the multivariate case, they are certainly much easier to handle. The class of models is best illustrated with an example.

Example 5.2. A bivariate VAR(1) model. Suppose two time series, X_{1t} and X_{2t}, have been observed where each variable is thought to depend linearly on the values of both variables in the previous time period. Then a suitable model for the two time series could consist of the following two equations:

$$X_{1t} = \phi_{11} X_{1,t-1} + \phi_{12} X_{2,t-1} + Z_{1t} \qquad (5.3.2)$$

$$X_{2t} \;=\; \phi_{21} X_{1,t-1} + \phi_{22} X_{2,t-1} + Z_{2t} \tag{5.3.3}$$

where $\{\phi_{ij}\}$ are constants and the innovations Z_{1t} and Z_{2t} constitute bivariate white noise. This means that the innovations have zero mean, and are uncorrelated through time (both within and between series), though it is possible that values of Z_{1t} and Z_{2t} could be correlated at the same time point. The model can be rewritten in matrix form as

$$\boldsymbol{X}_t = \Phi_1 \boldsymbol{X}_{t-1} + \boldsymbol{Z}_t \tag{5.3.4}$$

where $\boldsymbol{Z}_t^T = (Z_{1t}, Z_{2t})$ and

$$\Phi_1 = \begin{bmatrix} \phi_{11} & \phi_{12} \\ \phi_{21} & \phi_{22} \end{bmatrix}$$

Since $\boldsymbol{X}_t$ depends on $\boldsymbol{X}_{t-1}$, it is natural to call it a *vector autoregressive model* of order 1, abbreviated VAR(1). Equation (5.3.4) can be further rewritten as

$$(I - \Phi_1 B)\boldsymbol{X}_t = \boldsymbol{Z}_t \tag{5.3.5}$$

where B denotes the backward shift operator, I is the (2×2) identity matrix, and $\Phi_1 B$ represents the operator matrix $\begin{bmatrix} \phi_{11}B & \phi_{12}B \\ \phi_{21}B & \phi_{22}B \end{bmatrix}$. Then $\Phi(B) = (I - \Phi_1 B)$ is a matrix polynomial of order one in B when written in the general form of a VARMA model as in (5.3.1). □

More generally a VAR model of order p, abbreviated VAR(p), can be written in the form

$$\Phi(B)\boldsymbol{X}_t = \boldsymbol{Z}_t \tag{5.3.6}$$

where $\boldsymbol{X}_t$ is an $(m \times 1)$ vector of observed variables, $\boldsymbol{Z}_t$ denotes multivariate white noise, and Φ is a matrix polynomial of order p in the backward shift operator B.

Looking back at Example 5.2, we notice that if ϕ_{12} is zero in (5.3.2), then X_{1t} does not depend on lagged values of X_{2t}. This means that, while X_{2t} depends on X_{1t}, there is no feedback from X_{2t} to X_{1t}. Put another way, this means any causality goes in one direction only, namely from X_{1t} to X_{2t}.[5] Thus, in this case, the two equations constitute an open-loop transfer function model as defined in Section 5.2.2 (see also Reinsel, 1997, Example 2.4). The second of the two equations in the VAR-form of the model, namely (5.3.3), corresponds to (5.2.9) on realizing that the second component of $\boldsymbol{X}_t$, namely X_{2t}, is now the output, Y_t, in (5.2.9), with $\delta(B) = (1 - \phi_{22}B)$, $d = 1$ and $\omega(B) = \phi_{21}$. The first of the two equations in the VAR-form of model, namely (5.3.2), corresponds to a univariate ARMA model for the input X_{1t} as in (5.2.16) with $\phi_x(B) = (1 - \phi_{11}B))$ and $\theta_x(B)$ equal to unity.

More generally, if a set of m variables, which follow a VAR(1) model,

---

[5] In economic jargon, we say there is no Granger-causality from X_{2t} to X_{1t} – see Lütkepohl, 1993, Section 2.3.1.

can be ordered in such a way that Φ_1 is lower triangular (meaning that all coefficients above the diagonal are zero), then the model in (5.3.4) becomes a series of unidirectional transfer function equations, each involving one more variable than the previous equation, with the last variable, X_{mt}, being the final output. More generally again, if m variables follow a VAR(p) or VARMA(p,q) model, then there is unidirectional causality if the $\{\Phi_i\}$ and $\{\Theta_i\}$ matrices can be arranged so that they are all lower triangular.

Notice that the ordering of the variables in the above discussion is somewhat arbitrary. In particular, if ϕ_{21} in (5.3.3) is zero, rather than ϕ_{12} in (5.3.2), then we get a transfer-function model in which the variable X_{1t} becomes the output while X_{2t} becomes the input. In this case the coefficient matrices become upper (rather than lower) triangular.

In contrast to the open-loop situation, a closed-loop system has 'outputs' which feed back to affect the 'inputs', and then the $\{\Phi_i\}$ matrices will not be triangular. Then we have a set of mutually dependent variables for which the general VAR model may well be appropriate.

The definition of a VAR model in (5.3.6) does not attempt to account for features such as trend and seasonality. Some authors add appropriate deterministic terms to the right-hand side of the equation to account for a constant term, for trend and for seasonality. For example, seasonality can be handled by using seasonal dummy variables. Other people prefer alternative approaches. For example, the data could be deseasonalized before modelling the data, especially if the aim is to produce seasonally adjusted figures and forecasts. Another way of treating seasonality is to incorporate it into the VAR model by allowing the maximum lag length to equal the length of the seasonal cycle, but this may increase the number of model parameters to an uncomfortable level. Differencing may also be employed, with seasonal differencing used to remove seasonality and first differencing used to remove trend. However, the use of differencing is also not without problems, particularly if *cointegration* is present (see Section 5.4).

5.3.4 VMA, VARIMA and VARMAX models

Vector moving average (VMA) models arise when $\Phi(B) = I$ and $p = 0$ in (5.3.1). It is rather unusual to use VMA models in practice, but they have some theoretical interest arising from the multivariate generalization of Wold's theorem which effectively says that any purely nondeterministic process can be represented as a VMA process of infinite order. A VARMA process can then be seen as a rational approximation to this infinite VMA.

Various further generalizations of VARMA models can readily be made. If $\Phi(B)$ includes a factor of the form $I(1 - B)$, then the model acts on the first differences of the components of $\boldsymbol{X}_t$. By analogy with the acronym ARIMA, such a model is called a vector ARIMA or VARIMA model. However, it should be emphasized that different series may require different orders of differencing in order to make them stationary and so it may not be appropriate in practice to difference each component of $\boldsymbol{X}_t$ in the same way. Thus differencing is not, in general, a satisfactory way of fitting VAR models to non-stationary data. Section 5.4 considers an alternative approach, based on the possible presence of a phenomenon called *cointegration*, which needs to be considered before differencing multivariate data.

VARMA models can be further generalized by adding terms involving additional exogenous variables to the right-hand side of (5.3.1) and such a model is sometimes abbreviated as a VARMAX model (X stands for eXogenous). Further details are given by Reinsel (1997, Chapter 8). A specialist reference on the aggregation of VARMA models is Lütkepohl (1987).

5.3.5 Fitting VAR and VARMA models

There are various approaches to the identification of VARMA models which involve assessing the orders p and q, estimating the parameter matrices in (5.3.1) and estimating the variance-covariance matrix of the 'noise' components. A variety of software has become available and this section does not attempt to give details of all these methods. Rather the reader is referred to Priestley (1981) for an introduction, to Watson (1994) for a brief review, and to Lütkepohl (1993) for a thorough treatment. Reinsel (1997) covers material similar to the latter book but in a somewhat terser style. It is interesting to note that, although VARMA models are a natural generalization of univariate ARMA models, they are not covered by Box et al. (1994).

The identification of VAR and VARMA models is inevitably a difficult and complicated process because of the large number of model parameters which may need to be estimated. The number of parameters increases quadratically with m and can become uncomfortably large when a VARMA(p, q) model has p or q larger than two, even with only two or three variables. This suggests that some constraints need to be placed on the model and *restricted* VAR modelling is the term used when some coefficients are set equal to zero. This may be done using external knowledge, perhaps allied with a preliminary data analysis, to identify coefficients which can, a priori, be taken to be zero. The zeroes often arise in blocks, and a matrix, where most of the parameters are zero, is called a *sparse* matrix. However, even with such restrictions, VARMA models are still quite difficult to fit, even with only two or three explanatory variables, and most analysts restrict attention to vector AR (VAR) models as a, hopefully adequate, approximation to VARMA models. Even then, there is still a danger of over-fitting and VAR models do not appear to provide as parsimonious an approximation to real-life multivariate data as AR models do for univariate data. Thus many analysts restrict attention even further to low-order VAR models.

The latter suggestion lies behind a technique called *Bayesian vector autoregression* (abbreviated BVAR) which is increasingly used to fit

VAR models. The approach can be used whether or not the analyst is a Bayesian. The technique essentially aims to prevent over-fitting by shrinking parameters higher than first order towards zero. The usual prior that is used for the parameters, called the *Minnesota prior*, has mean values which are consistent a priori with every series being a random walk. Other priors have also been tried (e.g. Kadiyala and Karlsson, 1993). A tutorial paper showing how to select an appropriate BVAR model is given by Spencer (1993).

One important tool in VARMA model identification is the correlation matrix function (see Section 5.1.3). The diagonal elements of the cross-correlation matrices at different lags are, of course, all unity and therefore contain no information about the process. It is the off-diagonal elements which are the cross-correlation coefficients and which, in theory at least, should be helpful in model identification. In practice, we have already seen in Section 5.2.2 that the cross-correlation function of two time series can be very difficult to interpret, because, as previously noted, the standard error of a cross-correlation is affected by the possible presence of autocorrelation within the individual series and by the possible presence of feedback between the series. Indeed the author readily admits that he has typically found it difficult to utilize cross-correlation (and cross-spectral) estimates in model identification. The analysis of three or more series is in theory a natural extension of bivariate data analysis, but in practice can be much more difficult and is best attempted by analysts with substantial experience in univariate and bivariate ARIMA model building.

In transfer function modelling (see Section 5.2.2), the problems with interpreting cross-correlations were (partially) overcome by prewhitening the input before computing the cross-correlations so as to estimate the impulse response function. As regards VARMA model building, an unresolved general question is the extent to which the component series should be filtered or prewhitened before looking at cross-correlations. Another unresolved question is whether and when it is better to difference data before trying to fit VAR or VARMA models. There is conflicting evidence on this point, some of which can be found in a special issue of the *Journal of Forecasting* (1995, No. 3) which is entirely devoted to VAR modelling and forecasting and gives many more references on this and other aspects of VAR modelling. Research on fitting VAR and VARMA models is ongoing.

5.3.6 Forecasting with VAR, VARMA and VARIMA models

Minimum mean square error (MMSE) forecasts can readily be computed for VAR, VARMA and VARIMA models by a natural extension of methods employed for univariate and transfer function models (see especially the rules in Section 5.2.2 for computing forecasts for transfer function models). Generally speaking, future values of the white noise are replaced with zeroes while future values of X_t are replaced with MMSE forecasts. Present and past values of X_t and of Z_t are replaced by the observed values and the (one-step-ahead forecast) residuals, respectively. The procedure is best illustrated by examples.

Example 5.3. Forecasting a VAR(1) model. For simplicity, suppose $m = 2$ and that $X_t^T = (X_{1t}, X_{2t})$ follows a VAR(1) process which may be written

$$X_t = \Phi_1 X_{t-1} + Z_t$$

where Φ_1 is a 2×2 matrix. Given data up to time N, find the (h-steps-ahead) forecast of X_{N+h}, for $h = 1, 2, \ldots$

Using the model equation, the best one-step-ahead forecast is given by $\hat{X}_N(1) = \Phi_1 X_N$, as the best forecast of Z_{N+1} is zero. In order to forecast two steps ahead, the model equation suggests using $\hat{X}_N(2) = \Phi_1 X_{N+1}$. However, since X_{N+1} is unknown at time N, it is replaced by its forecast, giving $\hat{X}_N(2) = \Phi_1 \hat{X}_N(1) = \Phi_1^2 X_N$, where $\Phi_1^2 = \Phi_1 \Phi_1$ is, of course, a 2×2 matrix found by matrix multiplication. More generally, we find $\hat{X}_N(h) = \Phi_1^h X_N$, and the reader will recognize that this has the same form as the forecast for a univariate AR(1) model except that scalar multiplication changes to a series of matrix multiplications.   □

Example 5.4. Forecasting a VARMA(1, 1) model. Suppose we have data up to time N for a VARMA(1, 1) model, which may be denoted by

$$X_t = \Phi_1 X_{t-1} + Z_t + \Theta_1 Z_{t-1}$$

Then applying the procedure described above, the MMSE forecasts at time N are given by

$$\hat{X}_N(1) = \Phi_1 X_N + \Theta_1 Z_N$$

and

$$\hat{X}_N(h) = \Phi_1^{h-1} \hat{X}_N(1)$$

for $h = 2, 3, \ldots$

Of course, these formulae assume complete knowledge of the model, including the values of the model parameters, and also assume that the white noise series is known exactly. In practice, the model parameters have to be estimated and the white noise process has to be inferred from the forecast errors (the residuals). For models like a VARMA(1, 1) model, that involve an MA term, it is necessary to generate the entire white noise sequence from the beginning of the series in order to get the latest value of $\hat{Z}_N$. This is done recursively using all the past data values together with appropriate starting values for the initial values of Z_t, namely $Z_1, Z_0, Z_{-1}, \ldots$). The latter are sometimes set to zero, or can themselves be estimated by *backcasting*, wherein the series is 'turned round' and estimated values of the white noise before the start of the series are 'forecast' in a backward direction. Details will not be given here.   □

Example 5.5. Forecasting with a VARIMA(0, 1, 1) model. By analogy with the notation in the univariate case, a VARIMA$(0, 1, 1)$ model is a model which, when differenced once, gives a VARMA$(0, 1)$ model, which is, of course, a VMA model of order 1. This may be written in the form

$$(I - B)\boldsymbol{X}_t = \boldsymbol{Z}_t + \Theta_1 \boldsymbol{Z}_{t-1}$$

The reader may note that this is the multivariate generalization of the ARIMA$(0, 1, 1)$ model for which *exponential smoothing* is optimal, and is therefore sometimes called the vector exponential smoothing model. The model may be rewritten in the form

$$\boldsymbol{X}_t = \boldsymbol{X}_{t-1} + \boldsymbol{Z}_t + \Theta_1 \boldsymbol{Z}_{t-1}$$

Given data up to time N, we may derive MMSE forecasts by applying the general procedure described above. We find

$$\hat{\boldsymbol{X}}_N(1) = \boldsymbol{X}_N + \Theta_1 \boldsymbol{Z}_N \tag{5.3.7}$$

and

$$\hat{\boldsymbol{X}}_N(h) = \hat{\boldsymbol{X}}_N(h - 1) = \hat{\boldsymbol{X}}_N(1)$$

for $h > 1$.

Equation (5.3.7) is the multivariate error-correction prediction formula for multivariate exponential smoothing corresponding to (4.3.3) for simple exponential smoothing – see Section 4.3.1. Equation (5.3.7) can also be rewritten as

$$\hat{\boldsymbol{X}}_N(1) = (I + \Theta_1)\boldsymbol{X}_N - \Theta_1 \hat{\boldsymbol{X}}_{N-1}(1)$$

which corresponds to the *recurrence* form of exponential smoothing – Equation (4.3.2).   □

Empirical evidence. Having seen how to find forecast formulae for VAR, VARMA and VARIMA models, the really important question is whether good forecasts result, meaning more accurate forecasts than alternative methods. The empirical evidence is reviewed in Section 6.4.2, where we find, as ever, that results are mixed. VAR models, for example, sometimes work well, but sometimes do not. The reader should note in particular that some studies give an unfair advantage to more complicated multivariate models, by assuming knowledge of future values of explanatory variables, so that forecasts are not made on an out-of-sample basis. The main motivation for VAR modelling often lies in trying to get a better understanding of a given system, rather than in trying to get better forecasts.

5.4 Cointegration

Modelling multivariate time series data is complicated by the presence of non-stationarity, particularly with economic data. One possible approach is to difference each series until it is stationary and then fit a vector ARMA model, thus effectively fitting a VARIMA model. However, this does not always lead to satisfactory results, particularly if different degrees of differencing are appropriate for different series or if the structure of the trend is of intrinsic interest in itself, particularly in regard to assessing whether the trend is deterministic or stochastic. An alternative promising approach, much used in econometrics, is to look for what is called *cointegration*.

As a simple example, we might find that X_{1t} and X_{2t} are both non-stationary but that a particular linear combination of the two variables, say $(X_{1t} - kX_{2t})$ *is* stationary. Then the two variables are said to be *cointegrated*. If we now build a model for these two variables, the constraint implied by the stationary linear combination $(X_{1t} - kX_{2t})$ needs to be incorporated into the model and the resulting forecasts, and may enable the analyst to avoid differencing the data before trying to fit a model.

A more general definition of cointegration is as follows. A (univariate) series $\{X_t\}$ is said to be *integrated of order d*, written $I(d)$, if it needs to be differenced d times to make it stationary. If two series $\{X_{1t}\}$ and $\{X_{2t}\}$ are both $I(d)$, then any linear combination of the two series will usually be $I(d)$ as well. However, if a linear combination exists for which the order of integration is less than d, say $I(d - b)$, then the two series are said to be cointegrated of order (d, b), written $\mathrm{CI}(d, b)$. If this linear combination can be written in the form $\boldsymbol{\alpha}^T \boldsymbol{X}_t$, where $\boldsymbol{X}_t^T = (X_{1t}, X_{2t})$, then the vector $\boldsymbol{\alpha}$ is called a *cointegrating vector*. If $\boldsymbol{\alpha}$ is a cointegrating vector, then so is $2\boldsymbol{\alpha}$ and $-\boldsymbol{\alpha}$, and so the size and sign of $\boldsymbol{\alpha}$ is arbitrary. It is conventional to set the first element of $\boldsymbol{\alpha}$ to be unity.

For the example given earlier in this section, where $(X_{1t} - kX_{2t})$ is stationary, suppose that X_{1t} and X_{2t} are both I(1). Then $d = b = 1$, $\boldsymbol{X}_t$ is CI(1, 1), and a cointegrating vector is $\boldsymbol{\alpha}^T = (1, -k)$.

In a non-stationary vector ARIMA model, there is nothing to constrain the individual series to 'move together' in some sense, yet the laws of economics suggest that there are bound to be long-run equilibrium forces which will prevent some economic series from drifting too far apart. This is where the notion of cointegration comes in. The constraint(s) implied by cointegration imply long-run relationships which enable the analyst to fit a more realistic multivariate model.

Error-correction models. A class of models with direct links to cointegration are the *error-correction models* (ECMs) of economists. Suppose, for example, that two variables, X_{1t} and X_{2t}, which are both non-stationary, are thought to have an equilibrium relationship of the form $(X_{1t} - kX_{2t}) = 0$. Then it is reasonable to suppose that any changes in the variables could depend on departures from this equilibrium. We can therefore write down a plausible model in the form

$$\nabla X_{1t} = (X_{1t} - X_{1,(t-1)}) = c_1(X_{1,(t-1)} - kX_{2,(t-1)}) + \varepsilon_{1t}$$

$$\nabla X_{2t} = (X_{2t} - X_{2,(t-1)}) = c_2(X_{1,(t-1)} - kX_{2,(t-1)}) + \varepsilon_{2t}$$

where c_1, c_2 are constants. Clearly we can rewrite these two equations to express X_{1t} and X_{2t} in terms of the values at time $(t-1)$, namely $X_{1,(t-1)}$ and $X_{2,(t-1)}$. The resulting model is a VAR(1) model which is non-stationary but for which a cointegrating relationship exists, namely $(X_{1t} - kX_{2t}) = 0$.

Early work on cointegration was often carried out within a regression-model format. After fitting a regression model to non-stationary series, the residuals were tested for stationarity. If the residuals were found to be stationary, then the linear relationship determined by the fitted regression model could be regarded as a cointegrating relationship. Nowadays tests for cointegration are usually carried out within a VAR-model framework using the above ECM form which also enables us to explore and estimate cointegrated models more efficiently.

Suppose, for example, that the m-vector $\boldsymbol{X}_t$ follows a VAR(2) model

$$\boldsymbol{X}_t = \Phi_1 \boldsymbol{X}_{t-1} + \Phi_2 \boldsymbol{X}_{t-2} + \boldsymbol{Z}_t \qquad (5.4.1)$$

where Φ_1 and Φ_2 are $(m \times m)$ matrices such that the process is non-stationary, and $\boldsymbol{Z}_t$ denotes an $(m \times 1)$ multivariate white-noise vector. This may be rewritten in the form

$$\Phi(B)\boldsymbol{X}_t = \boldsymbol{Z}_t \qquad (5.4.2)$$

where $\Phi(B) = I - \Phi_1 B - \Phi_2 B^2$ is a matrix polynomial of order two in the backward shift operator B and I denotes the $(m \times m)$ identity matrix. If some components of $\boldsymbol{X}_t$ are non-stationary, then it follows that the equation: determinant$\{\Phi(x)\} = |\Phi(x)| = 0$ has one or more roots on the unit circle, when $x = 1$. Now, when $x = 1$, $\Phi(x)$ becomes $\Phi(1) = (I - \Phi_1 - \Phi_2) = C$ say, where C is an $(m \times m)$ matrix of constants. (Note: Do not confuse $\Phi(1)$ with Φ_1.) Thus, if the process is non-stationary, the determinant of C must be zero, which in turn means that C is not of full rank. We denote the rank of C by r where the values of interest are $0 < r < m$. (If $r = m$, then C is of full rank and $\boldsymbol{X}_t$ is stationary. If $r = 0$, C reduces to the null matrix which is of no interest here as it means that there are no cointegrating relationships, but rather a VAR(1) model could be fitted to the series of first differences, namely $\nabla \boldsymbol{X}_t$.) We see below that the rank of C is crucial in understanding cointegrating relationships.

The VAR(2) process given in (5.4.1) was defined to be non-stationary, and this normally makes it difficult to identify and estimate the model directly. Thus it may seem natural to take first differences of every term in the equation. However, that would induce non-invertibility in the innovation process. Instead we simply subtract $\boldsymbol{X}_{t-1}$ from both sides of the equation to give

$$\nabla \boldsymbol{X}_t = \boldsymbol{X}_t - \boldsymbol{X}_{t-1} = (\Phi_1 - I)\boldsymbol{X}_{t-1} + \Phi_2 \boldsymbol{X}_{t-2} + \boldsymbol{Z}_t \qquad (5.4.3)$$

The right-hand side of this equation can be rewritten in two different ways to incorporate $\nabla \boldsymbol{X}_{t-1}$, namely

$$\nabla \boldsymbol{X}_t = -\Phi_2 \nabla \boldsymbol{X}_{t-1} - C \boldsymbol{X}_{t-1} + \boldsymbol{Z}_t \qquad (5.4.4)$$

and

$$\nabla \boldsymbol{X}_t = (\Phi_1 - I)\nabla \boldsymbol{X}_{t-1} - C\boldsymbol{X}_{t-2} + \boldsymbol{Z}_t \qquad (5.4.5)$$

Notice that the coefficient of $\boldsymbol{X}_{t-1}$ in (5.4.4) and the coefficient of $\boldsymbol{X}_{t-2}$ in (5.4.5) are both equal to $C = (I - \Phi_1 - \Phi_2)$ as defined earlier in this section. Both these equations are of ECM form. If the first differences are stationary, and the innovations are stationary, then it follows that $C\boldsymbol{X}_{t-1}$ or $C\boldsymbol{X}_{t-2}$ will also be stationary and this explains why the matrix C is the key to determining any cointegrating relationships that exist. When C has rank r, it turns out that there are r cointegrating relationships and, by writing C as a product of matrices of order $(m \times r)$ and $(r \times m)$ which are both of full rank r, it can be shown that the latter is the matrix consisting of the r cointegrating vectors. It turns out that it is usually easier to test for the order of cointegration, to estimate the resulting model parameters and to enforce the resulting cointegration constraints by employing an ECM model form such as that in (5.4.4) or (5.4.5).

There is a rapidly growing literature concerned with these procedures, such as *Johansen's test* for the order of cointegration (i.e. the number of cointegrating relationships). There is also guidance on what to do if a cointegrating relationship is suspected a priori, namely to form the series generated by the cointegrating relationship and then test this series for the presence of a unit root. If cointegration exists, the generated series will be stationary and the null hypothesis of a unit root will hopefully be rejected. Further details on cointegration are given, for example, by Banerjee et al. (1993), Dhrymes (1997), Engle and Granger (1991), Hamilton (1994, Chapters 19, 20), Johansen (1996) and Lütkepohl, 1993, Chapter 11). An amusing non-technical introduction to the concept of cointegration is given by Murray (1994). Note that another class of models which exhibit cointegration, called *common trend models*, will be introduced in Section 5.6.

Despite all the above activity, the literature has hitherto paid little explicit attention to forecasting in the presence of cointegration. Clements and Hendry (1998a, Chapter 6) and Lütkepohl (1993, Section 11.3.1) are exceptions. A key question is whether one wants to forecast the individual variables (the usual case) or a linear combination of variables that makes the system cointegrated. The individual variables will typically be non-stationary and will therefore generally have prediction intervals that get wider and wider as the forecasting horizon increases. In contrast a stationary linear combination of variables will generally have finite prediction intervals as the horizon increases. Lütkepohl (1993) shows that the minimum MSE forecast for the individual variables in a cointegrated VAR process is given by a conditional expectation of the same form as in the stationary case. Lütkepohl (1993) also demonstrates that, in cointegrated systems, the MSE of the h-step-ahead forecast error will be unbounded for some variables but could approach an upper bound for other variables, even though they may be non-stationary. Clements and Hendry (1998a) are particularly concerned with evaluating forecasts when the model may possibly be misspecified.

I have not seen much empirical evidence on forecasts from cointegrated models. In theory, the presence of cointegrating relationships should help to improve forecasts particularly at longer lead times. However, although this is true for simulated data, improved forecasts do not necessarily result for real data (Lin and Tsay, 1996). Christoffersen and Diebold (1998) found that cointegrated models gave more accurate forecasts at short horizons but (surprisingly?) not for longer horizons. This may be because the standard measures of forecast accuracy fail to value the maintenance of cointegrating relationships among variables. The results in Kulendran and King (1997) appear to suggest that ECMs do not always lead to improved forecasts at short horizons, but the models that they call ECMs do not appear to include any cointegrating relationships but rather restrict attention to models for first differences.

I recommend that cointegration should always be considered when attempting to model and understand multivariate economic data, although there is, as yet, little evidence on whether the use of cointegration will lead to improved out-of-sample forecasts.

5.5 Econometric models

The term *econometric model* has no consistent usage but in this section is used to describe the large-scale models built by econometricians to reflect known theory about the behaviour of the economy and the relationships between economic variables. Given that economic data are influenced by policy decisions which involve 'feedback' from output variables to those input variables that can be controlled, it is clear that a single-equation model will be inadequate. Thus a sensible model for economic data should comprise more than one equation, and will typically contain tens, or even hundreds, of simultaneous equations.

There are some similarities with VAR models, which also comprise more than one equation, but there may also be fundamental differences, not only in the number of equations (econometric models are typically much 'larger') but also in the structure of the model and in the way that they are constructed. In econometric modelling, the variables are typically divided into endogenous and exogenous variables (see Section 5.3) where the latter may affect the system but are not affected by it. There is one equation for each endogenous variable. Econometric models allow exogenous variables to be included and so are more akin to VARMAX models (see Section 5.3.4). Econometric models also allow contemporaneous measurements on both the endogenous and exogenous variables to be included on the right-hand side of the equations.

As a simple example, suppose there are just two endogenous variables, say $Y_{1,t}$ and $Y_{2,t}$, and one exogenous variable, say x_t. Then we need two simultaneous equations to describe the system. A general way of representing the system will include both current and lagged values of the endogenous and exogenous variables and may be written as

$$
\begin{aligned}
Y_{1,t} \;=\;& aY_{2,t} + f_1(\text{lagged values of } Y_{1,t} \text{ and } Y_{2,t}) \\
& + bx_t + g_1(\text{lagged values of } x_t) + u_{1,t} \qquad (5.5.1)
\end{aligned}
$$

$$
\begin{aligned}
Y_{2,t} \;=\;& cY_{1,t} + f_2(\text{lagged values of } Y_{1,t} \text{ and } Y_{2,t}) \\
& + dx_t + g_2(\text{lagged values of } x_t) + u_{2,t} \qquad (5.5.2)
\end{aligned}
$$

where a, b, c, d are constants (some of which could be zero), f_1, f_2, g_1, g_2 are assumed to be linear functions and $u_{1,t}, u_{2,t}$ are error terms which are often assumed to be uncorrelated, both individually through time and with each other. This pair of equations is sometimes called the *structural* form of the model. If terms involving the current values of $Y_{1,t}$ and $Y_{2,t}$ are taken over to the left-hand side, and the pair of equations are written in matrix form, then the left-hand side of (5.5.1) and (5.5.2) may be written as $\begin{pmatrix} 1 & -a \\ -c & 1 \end{pmatrix} \begin{pmatrix} Y_{1,t} \\ Y_{2,t} \end{pmatrix}$ or $\Gamma \mathbf{Y}_t$ say, where Γ is the requisite (2×2) matrix and $\mathbf{Y}_t^T = (Y_{1,t}, Y_{2,t})$. If we now multiply through the matrix form of the structural model equation on the left by the inverse of Γ, then we obtain what is called the *reduced form* of the model in which the left-hand side of each equation is a single endogenous variable, while the right-hand side of the equations do NOT involve current values of the endogenous variables. The reduced form of the model is often easier to handle for the purpose of making forecasts. The variables on the right-hand side of the equations are now what economists call the *predetermined variables*, in that they are determined outside the model. However, the term is something of a misnomer in that current values of exogenous variables may be included, and the value for time $(N+1)$ may not be known at time N when a forecast is to be made. Partly for this reason, econometric models are often used for making *ex-post* forecasts, rather than *ex-ante* forecasts.

Three other differences from VAR models should be noted. First, there is no obvious reason why an econometric model should be restricted to a linear form (as in the mathematical discussion above), and non-linear relationships are often incorporated, especially in regard to the exogenous variables. Secondly, econometric modelling typically gives less attention to identifying the 'error' structure than statistical modelling, and may make assumptions that are overly simplistic. Thirdly, and perhaps most importantly, econometric models are generally constructed using economic theory rather than being identified from data.

Following up this last point, the contrasting skills and viewpoints of econometricians and statisticians were briefly mentioned earlier in comments on modelling in Section 3.5 and elsewhere. While econometricians tend to rely on economic theory, statisticians tend to rely on the data. As is usually the case in such situations, a balanced middle way, which utilizes the complementary nature of the skills used by both groups, may well be 'best'. It is certainly true that models based on economic theory need to be validated with real data. On the other hand, a statistical data-based approach will not get very far by itself without some economic guidelines for two reasons. First, an unrestricted analysis will find it difficult to select a sensible model from a virtually infinite number of choices, given that an economy may have a complex, non-linear structure which may be changing through time. Second, the nature of economic data makes it difficult to use for making inferences. Data sets are often quite small and will have other drawbacks, such as being collected while policy decisions are unclear or changing through time. Klein (1988) discusses the non-experimental nature of economics and the (lack of) precision in much economic data. In a time-series text, it is not appropriate to say more about this topic, except to say that econometric modelling is an iterative procedure which should normally involve both theory and data. Some further comments may be found in Granger and Newbold (1986, Section 6.3).

By including relevant policy variables, econometric models may be used to evaluate alternative economic strategies, and improve understanding of the system (the economy). These objectives are outside the scope of this book. Rather we are concerned with whether econometric models can also be used to produce forecasts. Although forecasting may not be the prime objective of econometric modelling, it does occur but in a rather different way to other forms of time-series forecasting. It is often necessary to use future values of exogenous variables, which can be done by forecasting them, by using the true values or by making assumptions about possible future scenarios. Thus forecasts may be either of the 'What-if' or of the *ex-post* variety, rather than being *ex-ante* or genuine out-of-sample. Moreover, despite their complexity, econometric forecasts are often judgmentally adjusted. For example, Clements and Hendry (1996, p. 101) say that econometric "forecasts usually represent a compromise between the model's output and the intuition and experience of the modeller". Even so, there is not much evidence that econometric models provide better forecasts than alternative time-series approaches – see Chapter 6. It is particularly galling to economists that univariate time-series models, which neither explain what is going on nor give any economic insight, may give better out-of-sample forecasts than econometric models. This may be because univariate models are more robust to structural breaks – see Section 8.6.3. Thus, despite their importance to economists, it seems that econometric models are often of limited value to forecasters.

5.6 Other approaches

Multivariate state-space (or structural) models have also been investigated (Harvey, 1989, especially Chapters 7 and 8). By analogy with Section 5.2, we can generalize the single-equation state-space model introduced in Section 3.2 by adding explanatory variables to the right-hand side of the

equation. For example, the observation equation of the linear growth model
may be generalized by adding a linear term in an explanatory variable. In
keeping with the notation in this chapter, we denote the response variable
by Y_t and the explanatory variable by x_t, but we denote the regression
coefficient by ν, rather than β, to avoid confusion with the notation for the
growth rate, β_t, which is the change in the current level, μ_t, at time t. The
observation equation can then be written as

$$Y_t = \mu_t + \nu x_t + n_t \qquad (5.6.1)$$

where $\{n_t\}$ denotes the observation errors which are usually assumed to be
white noise. The two transition equations for updating the level, μ_t, and
the growth rate, β_t, can be written in the same form as in the linear growth
model, namely

$$\mu_t = \mu_{t-1} + \beta_{t-1} + w_{1,t}$$
$$\beta_t = \beta_{t-1} + w_{2,t}$$

where $\{w_{1,t}\}$ and $\{w_{2,t}\}$ denote white noise processes, independent of $\{n_t\}$,
as in Section 3.2. Some advantages of trying to express regression-type
terms within the context of a state-space model is that the model can
be handled using state-space methods, there is no need to difference the
variables to make them stationary before commencing the analysis, and
there is no need to assume that all model parameters are unchanging
through time. While inference about the 'regression' parameter ν may be
of prime concern, the model does allow the analyst to estimate local values
of the level and trend in an adaptive way which does not force the analyst
to make inappropriate assumptions. Using a regression model, it is more
usual to impose a deterministic time trend which could, of course, give
misleading results if an alternative model for the trend is appropriate.

By analogy with Section 5.3, we can further generalize the single-equation
state-space models to the case where there is more than one 'response'
variable and so more than one observation equation. In keeping with the
notation in this chapter, we denote the $(m \times 1)$ vector of observed variables
by $\boldsymbol{X}_t$. We could, for example, generalize the random walk plus noise model
of Section 2.5.5 by considering the model

$$\boldsymbol{X}_t = \boldsymbol{\mu}_t + \boldsymbol{n}_t \qquad (5.6.2)$$

where

$$\boldsymbol{\mu}_t = \boldsymbol{\mu}_{t-1} + \boldsymbol{w}_t \qquad (5.6.3)$$

denotes the current mean level at time t, and $\boldsymbol{n}_t$ and $\boldsymbol{w}_t$ denote vectors
of disturbances of appropriate length. It is usually assumed that all the
components of $\boldsymbol{n}_t$ are uncorrelated with all components of $\boldsymbol{w}_t$, but that
the components of $\boldsymbol{n}_t$ (and perhaps of $\boldsymbol{w}_t$) may be correlated. The model
implies that the observed series are not causally related but, when applied
to economic data, allows for the possibility that the series are affected
by the same market conditions and so tend to move together because the
observation errors (the elements of $\boldsymbol{n}_t$) *are* correlated. The set of equations

constituting the above model are sometimes called a system of *Seemingly Unrelated Time Series Equations* – abbreviated SUTSE.

A variation on the SUTSE model is to include a vector of *common trends* which may affect all of the response variables. Then (5.6.2) is revised to become

$$\boldsymbol{X}_t = \Theta \boldsymbol{\mu}_t + \boldsymbol{\mu}_t^* + \boldsymbol{n}_t \qquad (5.6.4)$$

where Θ is a $(m \times r)$ matrix of constants, sometimes called factor loadings, $\boldsymbol{\mu}_t$ is the $(r \times 1)$ vector of common trends which is updated as in (5.6.3), and $\boldsymbol{\mu}_t^*$ is an $(m \times 1)$ vector of constants for which the first r elements are zero. The elements of $\boldsymbol{\mu}_t$ may affect all the elements of $\boldsymbol{X}_t$, whereas the non-zero elements of $\boldsymbol{\mu}_t^*$ only affect the corresponding element of X_t. Restrictions need to be placed on the values of Θ and on the covariance matrix of $\boldsymbol{w}_t$ in order to make the model identifiable. Because of the common trends, it turns out that the model is cointegrated. There are in fact $(m - r)$ cointegrating vectors.

A simple example may help to demonstrate this. Suppose $m = 2$, $r = 1$, so that Θ is a (2×1) vector. In order to make the model identifiable, it is convenient to set the first element of Θ to unity and write the model as

$$X_{1t} = \mu_t + 0 + n_{1t}$$

$$X_{2t} = \theta \mu_t + \bar{\mu} + n_{2t}$$

$$\mu_t = \mu_{t-1} + w_{1t}$$

Thus we take $\Theta^T = (1, \theta)$ and $\mu^{*T} = (0, \bar{\mu})$. As μ_t follows a random walk, it is clearly non-stationary. As X_{1t} and X_{2t} both depend on μ_t, they will also be non-stationary. However, if we use simple algebra to eliminate μ_t, it can easily be shown that the linear combination

$$X_{2t} - \theta X_{1t} = \bar{\mu} - \theta n_{1t} + n_{2t}$$

is a stationary process. Thus X_{1t} and X_{2t} are cointegrated and, in the notation of Section 5.4, we say that $\boldsymbol{X}_t$ is CI(1,1) with cointegrating vector given by $\boldsymbol{\alpha}^T = (-\theta, 1)$. The latter can readily be rewritten into the more conventional form, where the first element of $\boldsymbol{\alpha}$ is unity, as $\boldsymbol{\alpha}^T = (1, -1/\theta)$.

Multivariate Bayesian forecasting has been fully described by West and Harrison (1997, especially Chapters 9 and 16). Models with a single observation equation incorporating regression type terms can be handled by a single-equation dynamic linear model (DLM), while the multivariate DLM can be used to describe data with several response variables.

Intervention analysis aims to model the effect of one-off external events, such as might arise due to the effect of a labour dispute on sales, or of a change in the law on criminal statistics. A full account is given by Box et al. (1994, Chapter 12). The basic idea is to suppose that an intervention is known to occur in time period τ say, and to model its effect with a dummy variable, I_t say, which could have one of several forms. For example, if the intervention has an immediate effect for one time period only then we take

I_t to be of the form

$$I_t = \left\{ \begin{array}{ll} 1 & t = \tau \\ 0 & t \neq \tau \end{array} \right.$$

This is sometimes called a *pulse* function and can be recognized as a type of indicator variable. Alternatively, if the intervention causes a permanent shift in the response variable, then we take I_t to be of the form

$$I_t = \left\{ \begin{array}{ll} 0 & t < \tau \\ 1 & t \geq \tau \end{array} \right.$$

This is a *step change* type of variable. Other forms of I_t are also possible.

A simple example of a model for a response variable, Y_t, incorporating an intervention term is the following linear regression model, which, for simplicity, assumes no lag structure:

$$Y_t = \alpha + \beta x_t + \gamma I_t + \varepsilon_t$$

Here x_t denotes an explanatory variable, α, β and γ are constants and the intervention effect, I_t, is assumed to be of a known form such as one of those given above. Thus the time the intervention occurs, namely τ, is also a parameter of the model. The value of γ, as well as the values of α and β, can readily be estimated, for example, by least squares or maximum likelihood. Algebraically the indicator variable can be treated like any other explanatory variable and intervention variables can be incorporated into many other types of time-series model such as structural models (Harvey, 1989). The presence of an intervention is usually indicated by external knowledge, perhaps supplemented, or even instigated, by seeing an outlier or a step change or some other change in character in the observed time series. It is customary to assume that the value of τ, the time when the intervention occurs, is known from external knowledge, though it is possible to estimate it if necessary.

The use of *multivariate non-linear* time-series models for forecasting is still in its infancy and will not be discussed in detail here. Granger and Teräsvirta (1993) and Teräsvirta et al. (1994) introduce various types of multivariate non-linear model and discuss how they may be applied to economic data. Tong (1990, p. 101) introduces two generalizations of the threshold model to incorporate an observable input under either open-loop or closed-loop conditions, and later (Tong, 1990, Section 7.4) presents some examples on riverflow with precipitation and temperature as explanatory variables. As another example of the many variants of a multivariate nonlinear model which can be introduced, Niu (1996) proposes a class of additive models for environmental time series where with a response variable, Y_t, and p explanatory variables $x_{1t}, x_{2t}, \ldots, x_{pt}$, the model for Y_t

is of the (nonlinear) generalized additive form[6]

$$Y_t = f_0(t) + \sum_{j=1}^{p} f_j(x_{jt}) + \xi_t$$

where $f_0(t)$ denotes seasonal pattern and/or trend, the $f_j(.)$'s are arbitrary univariate functions and the noise series $\{\xi_t\}$ is described by an ARMA process formed from a white noise process $\{Z_t\}$ whose variance is not constant but depends exponentially on the explanatory variables in the form

$$\sigma_t^2 = exp\{\sum_{j=1}^{p} \beta_j x_{jt}\}$$

Thus both the mean level and variance of the process are modelled as nonlinear functions of relevant meteorological variables. The model has been successfully applied to environmental data giving more accurate estimates of the percentiles of the distribution of maximum ground-level ozone concentration. The focus is on with-sample fit rather than out-of-sample forecasting, but the scope for this, and many other classes of model, to be used in forecasting appear promising.

5.7 Some relationships between models

Section 4.4 showed that there are some close relationships between the many different linear models and forecasting methods in the univariate case. Likewise, in the multivariate case, we have already indicated some of the interrelationships that exist among the different classes of linear multivariate models, and this inevitably leads to connections between the resulting forecasts. This section briefly extends these ideas.

Consider, for example, a regression model, such as the leading indicator model in (5.1.1), namely

$$Y_t = a + bX_{t-d} + \varepsilon_t \tag{5.7.1}$$

For a regression model, it is customary to assume that the error terms, ε_t, are uncorrelated through time, even though this might be thought fairly unlikely! The above model may be regarded as a transfer function model, of the type given by (5.2.5), namely

$$Y_t^* = \nu_0 X_t + \nu_1 X_{t-1} + \ldots + n_t = \nu(B)X_t + n_t$$

by setting $Y_t^* = Y_t - a$, $\nu_j = b$ when $j = d$ and zero otherwise, and $n_t = \varepsilon_t$. Of course, if you really believe this model, there is not much point in expressing it as a transfer function model, as the forecast formulae will be identical. However, it does clarify the fact that a regression model is a very special, and arguably unlikely, case of the more general model, and,

---

[6] Note that Niu (1996) reverses the order of the subscripts so that X_{ti} is the t-th observation on x_i.

by fitting the more general transfer function model, the analyst can see if the estimated impulse response function (the $\{\hat{\nu}_i\}$) really does indicate a model of the form (5.7.1).

A transfer function model can, in turn, be expressed as a special type of VAR model by ensuring that the multivariate AR coefficient matrices are of triangular form, so that there is no feedback. For the simple regression model above, this can be achieved by something of a mathematical 'fiddle' in that one equation is a simple identity, namely $X_t = X_t$. Writing $Y_t^* = Y_t - a = bX_{t-d} + \varepsilon_t$, the corresponding VAR($d$) model can be expressed as

$$\left(\begin{array}{c} X_t \\ Y_t^* \end{array} \right) = \left(\begin{array}{cc} 1 & 0 \\ b & 0 \end{array} \right) B^d \left(\begin{array}{c} X_t \\ Y_t^* \end{array} \right) + \left(\begin{array}{c} 0 \\ \varepsilon_t \end{array} \right)$$

Clearly, there would generally be little point in expressing a regression model in the above VAR form, where one equation is just an identity. Thus the above example should be seen simply as a way of linking the different types of model and demonstrating the crude nature of regression time-series models, particularly in regard to the way that the 'error' terms are modelled. Allowing for autocorrelation between successive observation errors for the same variable, and allowing contemporaneous cross-correlations between the various observation errors for the different observed variables, is very important for setting up an adequate multivariate model.

A VAR model can, in turn, be expressed as a linear state space model – see, for example, Lütkepohl (1993, Section 13.2), but this representation will generally not be unique. A state space formulation could be helpful if we wish to make use of the Kalman filter to estimate the current 'state' of the system and make forecasts. However, the details will not be given here.

Earlier in the chapter, we noted the connection between error-correction models, common trend models and cointegrated VAR models, which should lead hopefully to a better understanding of the relationships that exist among these particular multivariate time series models, which are not easy to understand at first. More generally, the task of modelling relationships through time *and* between a set of variables, can make multivariate time-series modelling appear to be very complex. It is not for the faint-hearted, and should only be attempted by analysts with experience in analysing univariate time-series data as well as multivariate data that are not of time-series form. Indeed, in many situations there is much to be said for sticking to simpler forecasting methods and the reader who chooses to avoid the multivariate methods described in this chapter, will often find that little has been lost in terms of forecasting accuracy.

CHAPTER 6

A Comparative Assessment of Forecasting Methods

This chapter makes a comparative assessment of the many different methods and approaches which can be used to make forecasts. After a brief introduction, Section 6.2 considers criteria for comparing alternative forecasting methods, and Section 6.3 looks in particular at different ways of measuring forecast accuracy. Section 6.4 describes the different types of empirical study which have been carried out, including their general advantages and disadvantages, and then summarizes the main findings of empirical studies. Finally, Section 6.5 makes some suggestions on how to choose an appropriate forecasting method for a particular situation. The recommendations are based on research evidence up to the time of writing and update and expand earlier reviews by the author (Chatfield, 1988a; 1996a, Section 5.4; 1997).

6.1 Introduction

Chapters 4 and 5 introduced a wide variety of forecasting methods. It seems natural at this point to ask a question along the lines of:

"What is the best method of forecasting?"

This question, while superficially attractive, is in fact inadequately specified and admits no simple answer. The response can only be that *it depends on what is meant by 'best'*. Thus a method which is appropriate for forecasting one particular series in one particular context may be completely inappropriate for forecasting different data in a different context. A method appropriate for a single series may not be suitable for use with a large collection of series, especially when they have heterogeneous properties. A method used successfully by one analyst may be inappropriate for another forecaster, perhaps because he/she does not have the requisite knowledge or the necessary computer software.

When choosing a forecasting method, the reader is reminded (Section 1.2) that the first step in any forecasting exercise is to formulate the problem carefully by asking questions to clarify the context. As part of this exercise, the analyst will try to assess what is meant by 'best' when choosing a forecasting method. The next section therefore looks at the various criteria which need to be considered when choosing a forecasting method.

6.2 Criteria for choosing a forecasting method

The factors which need to be considered when choosing a forecasting method include the following:

(a) Forecast accuracy

(b) Cost

(c) Expertise of the analyst

(d) Availability of computer software

(e) The properties of the series being forecasted

(f) The way the forecast will be used

(g) Any other relevant contextual features

Much of the literature assumes (often implicitly) that 'best' means achieving the most accurate forecasts over the required time span, while ignoring other factors. This is unfortunate because accuracy is not always the overriding factor. A simple procedure, which is only marginally less accurate than a much more complicated one, will generally be preferred in practice. In any case, it is not always clear how accuracy should be measured and different measures of accuracy may yield different recommendations. However, forecast accuracy is a criterion which can be considered from a general statistical viewpoint and is the most obviously compelling to a statistician. Thus it is the only one considered in detail in this chapter – see Section 6.3. The other criteria, while important, are too context-dependent to say much about from a general methodological point of view.

6.3 Measuring forecast accuracy

As indicated above, we limit consideration of the 'best' forecasting method to mean the 'most accurate'. However, we still have the problem of deciding what is meant by 'most accurate'. This section discusses ways of measuring forecast accuracy, and hence of comparing forecasting results given by several different methods on the same data.

We have seen that the *evaluation* of forecasts is an important part of any forecasting exercise. Section 3.5.3 discussed various ways of checking the forecasts that arise from a *single* method, such as looking at the forecast bias (if any) and computing the residual autocorrelations (which should preferably be 'small'). Some of the ideas from these diagnostic checks, and from the related activity of *forecast monitoring*, are helpful in the problem we consider here, namely comparing the relative accuracy of several different forecasting methods on the same data. The diagnostic checks typically look at the *residuals*, namely the within-sample one-step-ahead forecast errors given by

$$e_t = x_t - \hat{x}_{t-1}(1) \tag{6.3.1}$$

as in Equation (3.5.1). The same formula is used to compute forecast errors made in out-of-sample mode, which we have seen is the preferred option. Clearly we want to make these forecast errors as small as possible (whatever that means) over the series as a whole.

Statisticians are used to measuring accuracy by computing *mean square error* (MSE), or its square root conventionally abbreviated by RMSE (for *root mean square error*). The latter is in the same units as the measured variable and so is a better descriptive statistic. However, the former is more amenable to theoretical analysis and that is where we begin.

When comparing forecasts from several different methods, the MSE form of statistic that is customarily calculated is the *prediction mean square error* (abbreviated PMSE), which is the average squared forecast error and is usually computed in the form

$$PMSE = \sum_{t=N-m+1}^{N} e_t^2 / m \qquad (6.3.2)$$

There are several aspects of (6.3.2) which deserve consideration. Firstly, statisticians often use MSE to measure accuracy without giving this choice much thought. In fact, the use of MSE implies an underlying quadratic *loss function* (or equivalently a quadratic cost function) and this is the assumed loss function for which the conditional mean is the 'best' forecast, meaning that it is the minimum MSE forecast (see Section 4.1). The concept of a *loss function* is a key component of forecasting but is often glossed over without much thought. It was introduced in Section 4.1 and attempts to quantify the loss (or cost) associated with forecast errors of various sizes and signs. For example, if the forecast error doubles, is this twice as bad (linear) or four times as bad (quadratic)? Is a positive error equally good (or bad) as a negative error of the same size? In other words, is the loss function symmetric? These and other questions need to be addressed in the *context* of the given situation and may depend partly on subjective judgement. Different users may on occasion prefer to choose different loss functions for the same problem, and this could lead to different models and different point forecasts.

Suppose, for example, that one analysts thinks the loss function is linear, rather than quadratic. Then the preferred estimate of accuracy is given by the *Mean Absolute Error* (abbreviated MAE), namely

$$MAE = \sum_{t=N-m+1}^{N} |e_t| / m \qquad (6.3.3)$$

Minimizing this measure of accuracy, rather than the MSE, will in general lead to different parameter estimates and could even lead to a different model structure.

Of course, if the modulus signs in (6.3.3) are omitted, then we have the *mean error* (abbreviated ME) given by

$$ME = \sum_{t=N-m+1}^{N} e_t/m \qquad (6.3.4)$$

which measures the overall forecast *bias* (if any), which is merely one component of the accuracy or precision. Note that it may be worth checking the bias for selected subsets of the data, as well as for all the data. For example, with economic data, it is worth checking to see if there is a bias of one sign during economic expansion and a bias of the opposite sign during economic contraction.

A second feature of (6.3.2) is that the sum is over the last m observations. When a comparative forecasting study is made, it is customary to split the data into two parts. The model is fitted to the first part of the data, and then forecasts are compared over the second part of the data, sometimes called the *test set*. If forecasts are to be genuine out-of-sample forecasts, as they should be, then this partitioning should be carefully carried out so that the test set is *not* used in fitting the model. Thus the model is fitted to the first $(N - m)$ observations and the last m observations constitute the test set.

A third feature of (6.3.2) is that all forecasts are one-step-ahead forecasts. Now when a time-series model is fitted by least squares, this will normally involve minimizing the sum of squared residuals, where the residuals are the *within-sample* one-step-ahead forecast errors. When assessing forecasts, it is best to look at *out-of-sample* forecast errors, and it is not immediately obvious whether all forecasts should be made from the same time origin at the end of the training set (and hence be at different lead times) or whether all forecasts be made for the same lead time (and hence be made from different time origins). Equation (6.3.2) effectively assumes that the observations in the test set are made available one at a time as successive forecasts are computed.

If forecasts are required h steps ahead, the question arises as to whether (6.3.2) should be modified in an obvious way to incorporate h-step-ahead forecasts and whether the model should be fitted by minimizing the within-sample h-step-ahead errors. If the analyst really believes that the true model has been found, then the answer to the last question is 'No'. In practice the analyst usually realizes that the fitted model is at best a good approximation and then it is advisable to focus on h-step-ahead errors for both fitting and forecasting if that is the horizon of interest (see Section 8.5).

The final feature of the PMSE statistic in (6.3.2), which needs highlighting, is that it depends on the scale in which the variable is measured. For example, if a series of temperatures are measured in degrees Fahrenheit, rather than Centigrade, then the PMSE will increase even if an identical forecasting procedure is used for both series. This means that, although PMSE may be appropriate for assessing the results for a single time series, it should not be used to assess accuracy *across* (many) different series, especially when the series have substantially different variances or are measuring different things on different scales. Failure to appreciate this point has resulted in misleading results being published (Chatfield, 1988b).

Various alternatives to MSE are available for assessing forecast accuracy either for a single series, or for averaging across series as in the sort of forecasting competition reviewed in Section 6.4. A good discussion is given by Armstrong and Collopy (1992) and Fildes (1992) and in the commentary which follows those papers. Clearly the measure of forecast accuracy should match the loss function thought to be appropriate. The user should also be clear whether the measure depends on the units in which the observations have been taken. In particular, if two or more forecasting methods are to be compared on more than one series, it is essential that any error measure, used for assessing accuracy *across* series, should be *scale-independent*. As noted above, this excludes the raw MSE.

There are several commonly used types of scale-independent statistic. The first type essentially relies on pairwise comparisons. If method A and method B, say, are tried on a number of different series, then it is possible to count the number of series where method A gives better forecasts than B (using any sensible measure of accuracy). Alternatively, each method can be compared with a standard method, such as the random walk forecast (where all forecasts equal the latest observation), and the number of times each method outperforms the standard is counted. Then the percentage number of times a method is better than a standard method can readily be found. This statistic is usually called '*Percent Better*'.

Another type of scale-independent statistic involves *percentage errors* rather than the raw errors. For example, the *mean absolute prediction error* (MAPE) is given by

$$MAPE = \sum_{t=N-m+1}^{N} |e_t/x_t|/m \qquad (6.3.5)$$

Statistics involving percentage errors only make sense for non-negative variables having a meaningful zero. Fortunately, this covers most variables measured in practice.

The so-called *Theil* coefficients, named after Henri Theil, are also scale-independent. There are (at least) two different Theil coefficients, which are often labelled as U_1 and U_2. For both coefficients, a value of zero indicates perfect forecasts. However, whereas U_1 is constrained to the interval (0,1), with values near unity indicating poor forecasts, the value of U_2 may exceed unity, and equals unity when it is no better or worse than the 'no-change' forecast. For this and other reasons, the coefficients are not easy to interpret. The values frequently appear on computer output, but few software users seem to understand them. Moreover, they cannot be recommended on theoretical grounds (Granger and Newbold, 1986, Section 9.3). Thus they will not be described here.

Even using a good scale-independent statistic, such as MAPE, it is possible that one series will give such poor forecasts using one or more of the methods being considered, that the results are dominated by this one series. For this reason, some sort of robust procedure is often used, either by trimming the set of data series (e.g. by removing one or more series which give much worse forecasts than the rest) or by using the *geometric* mean instead of the arithmetic mean when averaging across series.[1] Armstrong and Collopy (1992) and Fildes (1992) both recommend a scale-independent statistic based on the use of the geometric mean. However, my impression is that statistics involving the geometric mean are little used, partly because few people understand them and partly because computer software is not readily available to compute them. Note that the statistic 'Percent Better' is immune to outliers anyway.

Robust procedures can also be used when calculating measures of accuracy for a *single* series (rather than across series), if the distribution of forecast errors is clearly not normal. A few large errors, caused by sudden breaks or outliers in the time series, could dominate the remaining results. One approach is to downweight extreme errors (sometimes called *Winsorizing*) or remove them altogether. Another possibility is to use a geometric mean of the errors over the forecast period. Clements and Hendry (1998a, Section 3.6) have proposed a measure which they claim is invariant to non-singular, scale-preserving linear transformations (such as differencing), but this looks rather complicated to use in practice.

Finally it is worth noting that the use of standard statistical criteria, such as MAPE, which are chosen for their scale-independent properties, may not be consistent with alternative criteria such as those arising from economic considerations (Gerlow et al., 1993). For example, we noted earlier that the real loss function may be asymmetric (e.g. underforecasting is worse than overforecasting). Alternatively, it may be particularly important to forecast the *direction* of movement (will the next first difference be positive or negative?) or to predict large movements. As always, the use of standard statistical procedures may need to be modified as necessary within the context of the problem, although it may not be possible to select a measure of forecast accuracy that is scale-independent and yet satisfies the demands of the appropriate loss function. In a similar vein, the usual measures of forecast accuracy fail to value the maintenance of cointegrating relationships among the forecasted variables, when forecasting with a cointegrated model. Christoffersen and Diebold (1998) suggest alternatives that explicitly do so.

In conclusion, we can say that the search for a 'best' measure of forecast accuracy is likely to end in failure. There is no measure suitable for all types of data and all contexts. Moreover, there is plenty of empirical evidence that a method which is 'best' under one criterion need not be 'best' under

---

[1] The geometric mean of the values $s_1, s_2, \ldots, s_k$, is the product of the values raised to the power $1/k$.

alternative criteria (e.g. Swanson and White, 1997). However, it has been possible to make some general recommendations, notably that a measure of accuracy must be scale-independent if comparisons are made across series, and that the choice of measure must take account of the distribution of individual forecast errors for a single series, or the distribution of accuracy statistics from different series when averaging across a set of series, particularly in deciding when it is advisable to use a robust approach.

6.3.1 Testing comparative forecast accuracy

Whichever measure of forecast accuracy is selected, the analyst will often find that the results for different forecasting methods appear to be quite similar, so that there is no clear 'winner'. Rather, the 'best' method may be only a little better than alternatives as judged by the measure of forecast accuracy. Then the question arises as to whether there is a 'significant' difference between the methods. This is a rather hard question to answer as it is not immediately obvious what is meant by significance in this context. In asking if the differences between forecasting methods can be attributed to sampling variability, it is not clear what sampling mechanism is appropriate. Nor is it obvious what assumptions should be made about the different methods and about the properties of the forecast errors that arise from the different methods. Clements and Hendry (1998a, Section 13.3) describe some tests, but I would be reluctant to rely on them.

The real test of a method lies in its ability to produce out-of-sample forecasts. For a single series, the test set will often be quite small. Then significance tests will have poor power anyway. With a group of series, any assumptions needed for a significance test are likely be invalid and a robust nonparametric approach may be indicated. For example, if a method is better than alternatives for a clear majority of the series, then that is a good indication that it really is better. Practical significance is usually more interesting than theoretical significance and it will be often be a matter of judgement as to whether, for example, it is worth the extra effort involved in a more complicated method to improve a measure of accuracy by, say, 5%.

6.4 Forecasting competitions and case studies

Leaving theoretical issues to one side for the time being, the real test of a forecasting method is whether it produces good forecasts for real data. In other words, we want to know if a method works when applied in real-life situations.

Over the years, many investigations have been made in order to provide empirical evidence to try to answer this question. These empirical studies can loosely be divided into two types, usually called *forecasting competitions* and *case studies*, though there is no standard usage for these terms. This section explains how the two types of study are conducted, discusses their general advantages and disadvantages and then reviews the empirical evidence.

6.4.1 General remarks

A key feature of all empirical investigations is the number of series analysed. The term 'case study' is typically used when a small number of datasets are analysed, say between one and five. Such studies may also be called *empirical examples* or, when a new method has been proposed, an empirical *application*. In contrast, the term 'forecasting competition' is typically used when at least 25 datasets are analysed and perhaps many more. For example, the forecasting competition usually known as the *M-competition* (Makridakis et al., 1982, 1984) analysed 1001 series, which is a very large sample. However, there is no clear dividing line between these two types of study, and it is certainly true that 'bigger' is not necessarily 'better', as a smaller number of series can be analysed more carefully. Many studies use an intermediate number of datasets (e.g. between 10 and 50) and some authors use the term 'forecasting competition' even when the number of datasets is quite small. However, it will be convenient here to restrict the use of the term 'competition' to larger datasets.

In case studies, the analyst will typically be able to devote detailed attention to each dataset, apply the different forecasting methods carefully and take account of context. In contrast, forecasting competitions may analyse so many series that it is impractical for analysts to pay much attention to each series. In the M-competition, for example, it was decided to apply the Box-Jenkins method, not to all 1001 series, but to a systematic subsample of 111 series. Even so, it was reported that each series took over one hour to analyse and this enormous commitment can only rarely be made. Thus forecasting competitions, with large datasets, seem more suitable for assessing automatic univariate procedures, while case studies seem more suitable for assessing multivariate procedures and non-automatic univariate procedures.

Empirical studies can help to show whether the success or failure of particular forecasting methods depends on such factors as

(a) the type of data

(b) the length of the forecasting horizon

(c) the skill of the analyst in using the method (e.g. in selecting an appropriate model from the ARIMA class for Box-Jenkins forecasting) and

(d) the numerical methods used to fit a model or implement a method and compute predictions

Case studies are better suited for shedding light on factors such as (c) and (d), and few would dispute that case studies are an essential part of the development of any forecasting method. They may also shed some light on (a) and (b), but, by their restricted nature, make generalization problematic. Large-scale competitions are much more controversial. They seem particularly useful for shedding light on features such as (a) and (b) above, but mainly for simple automatic procedures (Chatfield, 1988a, Section 4.1). A historical review of the results of forecasting competitions is given in Section 6.4.2 but first their general advantages and disadvantages are assessed.

Over the years, many different forecasting competitions have been carried out, of which some were good and some not so good. By 'good' is meant that the data were carefully selected, that methods were applied competently and that the results were evaluated fairly and presented clearly. This has not always been the case, but even when it has, there have been wildly diverging views as to the value of competition results. Thus McLaughlin (Armstrong and Lusk, 1983, p. 274) described the M-competition (Makridakis et al., 1982, 1984) as 'a landmark for years to come', while Jenkins and McLeod (1982, preface) described forecasting competitions as being 'ludicrous'.

It will be helpful to examine the criticisms that have been levelled at competitions over the years.

(1) *The data*. It is impossible to take a random sample of time series, so that it is never clear to what extent the results from any competition can be generalized. One strategy is to take data from a variety of application areas, apply the different methods and hope the results will apply more generally. One danger with this is that one method may work well on data from one field of study but poorly on data from another area. There is, therefore, something to be said for restricting attention in any one competition to data from one particular area so that more specific conclusions can be made, while recognizing that the results may not generalize to other types of data. Case studies, by their nature, will often be small-scale versions of the latter type of study. This general criticism of the way that data are selected can be made of many competitions. There is no easy answer but it is surely the case that experience with particular datasets is generally better than nothing, even if the sample of datasets cannot be randomly selected.

(2) *The skill of the participants*. When method A outperforms method B in a competition, there can be no certainty that this necessarily implies that method A is better. It may be that method B has been applied imperfectly, especially when method B is of a complicated form. This argument has been used to criticize the way that some methods (notably Box-Jenkins) have been applied in competitions but it is possible to turn this argument on its head. The more complicated the method or model used to make forecasts, the more likely it is that it will be misapplied. Given that participants in competitions are likely to be (much) more competent than the average forecaster, it is arguable that competitions *overestimate* the likely accuracy of forecasts from complicated methods when applied in real-life situations. A similar argument would suggest that new methods are likely to have an

unfair advantage when applied by their originator. New methods need to be tried out by independent investigators.

(3) *Replicability.* It is a feature of good science that the results can be replicated by other researchers. This has not always been the case with forecasting competitions in the past, but it is increasingly true today and many journals now demand sufficient information about the methods and data to enable others to check the results. For example, the data from the M-competition are widely available and several researchers have carried out follow-up studies.

(4) *Accuracy criteria.* As noted in Section 6.3, there are many different ways of assessing the accuracy of forecasts. While there are obvious dangers in averaging results across different series which might have very different properties (and hence give very different percentage forecast errors), many competitions present more than one measure to summarize the results. Provided the measures are scale-independent, they should give some idea on relative accuracy. Of course, a method which is best by one criteria need not be best by another, and that is another indication that the search for a global best method is doomed to failure. In any case, accuracy is only one indicator of the 'goodness' of a forecast. Practitioners think cost, ease of use and ease of interpretation are of comparable importance (Carbone and Armstrong, 1982) but these other criteria are typically ignored in forecasting competitions.

(5) *Ignoring context?* Perhaps the most serious criticism of large-scale forecasting competitions is the objection to analysing large numbers of series in a predominantly automatic way, with no contact with a 'client' and no opportunity to take account of the context. Jenkins and Mcleod (1982) describe this as 'manifestly absurd' and go on to describe several one-off case studies where statistician and client collaborate to develop an appropriate multivariate model. The results of forecasting competitions would not help in such situations. However, there are some situations, such as stock (inventory) control, where large numbers of series are involved and it would be quite impractical to develop a model tailored to each series. Then a relatively simple, predominantly automatic, procedure is needed and the results of forecasting competitions are likely to be relevant. Clearly both automatic and non-automatic *approaches* to forecasting have their place, and the distinction between them is sometimes more important than the difference between forecasting *methods.*

I conclude that well-run forecasting competitions and case studies do provide some information, but that the results will only form part of the story. Competitions are mainly helpful for comparing predominantly automatic methods while case studies are better placed to tell us about non-automatic procedures. However, poorly managed studies could be worse than useless, and a key objective of this chapter is to encourage high standards when carrying out empirical studies. In particular, (i) data should be carefully chosen, (ii) forecasts should be made out of sample, (iii) good software should be used by an experienced analyst, (iv) more than one measure of accuracy should be tried, (v) sufficient information should be provided to make the results replicable.

6.4.2 Review of empirical evidence

This subsection reviews the results of the many empirical studies which have been carried out over the years to assess the forecasting accuracy of different forecasting methods on a variety of datasets. The review is necessarily selective and concentrates on the large-scale forecasting competitions which have been carried out since the early 1970's. Such competitions often restrict attention to comparing automatic univariate procedures, although some non-automatic methods (e.g. Box-Jenkins) have also been tried. Multivariate methods, and complicated univariate methods, have generally been investigated by means of case studies, and a brief summary of relevant results is given, though the results are often more concerned with shedding light on one particular method, rather than with making general comparisons. As such they may have been cited elsewhere in Chapters 4 and 5.

The review of empirical accuracy studies, given by Fildes and Makridakis (1995), has a rather different emphasis in that they are concerned to point out that empirical findings have often been ignored by the theoreticians. For example, sophisticated methods often fail to out-perform simpler methods, and yet the published methodology continues to emphasize more complicated methods. Likewise, out-of-sample forecasts are typically found to be much worse than expected from within-sample fit and yet many researchers continue to assume the existence of a known true model. In order to illustrate the above points, it is worth considering a very simple forecast procedure, namely that given by the random walk. Here the 'best' forecast of the next value is the same as the most recent value. This is a disarmingly simple method, but is often quite sensible, and has been widely applied to detrended, deseasonalized economic data even though many analysts expect that more complicated methods will generally be superior. Efforts to find a better way of predicting such series have often been unsuccessful. For example, Meese and Rogoff (1983) have shown empirically that the random walk will often give as good, or even better, forecasts *out of sample* than those given by various alternative univariate and multivariate models, while White (1994) found that the random walk often outperformed neural networks for economic data. Of course, alternative models may give a somewhat better *fit*, but this is often not a reliable guide to out-of-sample behaviour (see also Chapters 7 and 8).

We now attempt a summary of the results of forecasting competitions. As will be seen below, the results are not always consistent and may only be partially relevant to a new forecasting problem in a particular context. Nevertheless, the findings are (much) better than nothing and provide pointers for the choice of forecasting method, as will be clarified in Section 6.5.

Two early competitions were described by Newbold and Granger (1974) and Reid (1975). The first of these compared Box-Jenkins with Holt-Winters exponential smoothing, stepwise autoregression and various combinations of methods. They showed Box-Jenkins gave more accurate out-of-sample forecasts on average than the other two methods although Box-Jenkins required much more effort and could sometimes be beaten by a combination of methods. They recommended using Holt-Winters for short series (less than 30 observations) and Box-Jenkins for longer series. They also recommended considering the possibility of combining forecasts from different methods. In addition, they emphasized that any guidelines should not be followed blindly when additional information is available about a particular series and the recommendation to use contextual information wherever possible is one I would echo today. Reid's study used a larger selection of univariate methods and also found that Box-Jenkins was 'best', although its advantage decreased as the lead time increased.

The next major competition is the M-competition (Makridakis et al., 1982, 1984) which was mentioned at the start of Section 6.4.1. Compared with earlier studies, this was much larger, with more analysts (9), more methods (24) and more series (1001) of a very varied nature (see specimen graphs in Makridakis et al. 1984, pp. 281-287). The Box-Jenkins, ARARMA and FORSYS methods were applied to a systematic subsample of 111 series in view of their relative complexity. Various measures of accuracy were computed and tabulated, though the resulting tables are hard to 'read' (and the table of mean square errors should be disregarded as this statistic is not scale-independent – Chatfield, 1988b). Moreover, the interpretation of the results is unclear, perhaps because the nine analysts had difficulty in reaching a consensus. The reader is therefore recommended to read Armstrong and Lusk (1983) as well as Chapter 1 of Makridakis et al. (1984). The results were as expected in that no one method is best in every sense, but rather the choice of method depends on the type of data, the forecasting horizon, the way accuracy is measured and so on. What was perhaps not expected, given earlier results, is that Box-Jenkins did not give more accurate forecasts than several simpler methods.

The two most accurate methods, on the basis of the M-competition results, are the ARARMA method (Parzen, 1982) and FORSYS. The latter is a German proprietary method. However, both were implemented by their originators and, for the reasons explained in Chatfield (1998a, p. 30), the results should be treated with caution. This may explain why these two methods have not become more widely used, although there has been some subsequent interest in the ARAR algorithm (e.g. Brockwell and Davis, 1996, Section 9.1). There were four other methods which appeared to be at least as accurate as Box-Jenkins, namely deseasonalized Holt's exponential smoothing, Holt-Winters, Bayesian forecasting and a combination of six different methods called the combining A method. Of these methods,

Fildes (1983) says that "Bayesian forecasting is apparently not worth the extra complexity compared with other simpler methods", although later developments (e.g. Pole et al., 1994) may negate this comment. The combining A method is tedious and does not give an interpretable model. This leaves two types of exponential smoothing as general all-purpose automatic procedures and both can be recommended. In particular, the author has much experience with one of these methods, namely the Holt-Winters method. The latter is straightforward to implement and generally gives intuitively sensible results. Moreover, there is evidence that Holt-Winters is robust to departures from the model for which the method is optimal (Chen, 1997).

Recently, Hibon and Makridakis (1999) have reported the results of a follow-up to the M-competition, called the M3-competition, which analyses as many as 3003 series and 24 methods. The latter include neural networks and several expert systems. As in earlier studies, there was no clear winner but the results help to identify which methods are doing better than others for specific types of data, and for different horizons, though my impression of the results is that overall average differences are generally small.

There has been growing interest in *structural models* since the publication of Harvey (1989), though there has so far been rather little systematic empirical examination of the forecasts which result. The empirical study of Andrews (1994) is one exception and suggests the models do quite well in practice, especially for long horizons and seasonal data.

Competitions are much more difficult to organize for comparing non-automatic univariate methods and multivariate methods. Then smaller *case studies* may be more fruitful. This is illustrated by the M2-competition (Makridakis et al., 1993) which involved 29 time series and a variety of multivariate and non-automatic univariate methods as well as automatic univariate methods. Participants also received a variety of additional contextual information. I was one of the participants but found the exercise rather unsatisfactory in some ways, mainly because there was no direct contact with the 'client'. As a result, it was difficult to take full account of the additional information. Thus the conditions did not really mimic those likely to hold in practice when using such methods. Although called a competition, it is not clear whether a sample size of 29 series should make the M2-competition a large case study rather than a small forecasting competition, or neither. It is arguable that the number of series is too small to call it a competition but too large for a properly controlled case study. Put another way, the M2-competition suffers from using an 'in-between' sample size which is too large to allow proper attention to all the series, but too small to offer generalizable results. Thus the remainder of this subsection reports empirical results for assessing non-automatic univariate and multivariate methods, which really are of the smaller case study form. Given the difficulties in fitting such models, small case studies are more feasible and may well give more reliable results. We start with results for multivariate methods.

Results for multivariate models

Many forecasters expect multivariate forecasts to be at least as good as univariate forecasts, but this is not true either in theory or practice, for the reasons listed in Section 5.1. For example, multivariate models are more difficult to identify, and appear to be less robust to departures from the model than univariate models. An earlier review of empirical results is given by Chatfield (1988a). There have been some encouraging case studies using transfer function models (e.g. Jenkins, 1979, Jenkins and McLeod, 1982), but such models assume there is no feedback from the response variable to the explanatory variables and that will not apply in many situations (e.g. with most economic data). Econometric simultaneous equation models have a rather patchy record (e.g. Makridakis and Hibon, 1979, Section 2), not helped by the fact that economists often fail to compare their models with simpler alternatives and are usually more interested in description and explanation, rather than forecasting. Regression models also have a rather patchy record (Fildes, 1985).

The situation in regard to VAR models is more difficult to summarize and we devote more space to this topic. (Results for VARMA models are much less common and will not be reviewed here.) Many studies have been published in which forecasts from VAR models are compared with alternative methods. The results look promising in rather more than half the case studies that I have seen in such journals as the *International Journal of Forecasting*, the *Journal of Forecasting* (especially 1995, Part 3) and the *Journal of Business and Economic Statistics*. The reader is encouraged to browse in these and other journals. It is clear that the VAR models need to be carefully applied, but, even so, the results are not always consistent. Moreover, in some studies, the choice of alternative methods may seem inappropriate and designed to give an advantage to VAR models, while in others, insufficient information is given to enable the analyst to replicate the results. In addition, it is not always clear when VAR forecasts are genuine out-of-sample forecasts. I have seen several studies which claim that VAR forecasts are superior, but which are computed on an *ex-post* basis, meaning that future values of explanatory variables are assumed known. I have also seen examples where random walk forecasts are as good as VAR forecasts, and, with economic data, it is particularly important to ensure that out-of-sample VAR forecasts are at least superior to those given by the random walk model.

The literature is too large to cover thoroughly and so we simply refer to a couple of specimen studies. The results in Boero (1990) suggest that a Bayesian VAR model is better than a large-scale econometric model for short-term forecasting, but not for long-term forecasts where the econometric model can benefit from judgemental interventions by the model user and may be able to pick up non-linearities not captured by (linear) VAR models. As is often the case, different approaches and models are complementary. The results in Kadiyala and Karlsson (1993) suggest that unrestricted VAR models do not forecast as well as when using

Bayesian vector autoregression presumably because unrestricted models generally contain too many parameters and give a spuriously good fit (within sample), while giving out-of-sample forecasts which are poorer than alternatives.

Thus, although it can be fruitful, VAR modelling is not the universal panacea that some writers claimed it to be in the late 1980's (nor is any forecasting method!). To have a chance of success, it needs to be implemented carefully, using contextual information and subjective modelling skills, and the 'black-box' approach evident in some published case studies should be avoided.

The very limited empirical evidence on the use of cointegrated VAR models for forecasting was briefly reviewed in Section 5.4. No general recommendations can, as yet, be made.

Results for periodic and long-memory models
Turning to univariate models, we have already seen that there is a wealth of empirical evidence on the use of ARIMA (and SARIMA) models in forecasting. However, there is much less evidence to date on the many variants of ARIMA models, such as periodic and fractionally integrated ARMA models

Periodic AR and ARMA models have been fitted to economic data (Novales and Flores de Frato, 1997) and to riverflow data (McLeod, 1993). They suffer the usual danger of more complicated models that they may give an improvement in fit, as compared with simpler models, but do not necessarily give better out-of-sample forecasts. Moreover, it appears that about 25 years data are needed in order to have a reasonable chance of success in detecting changing seasonal behaviour and in estimating such effects reasonably accurately. This length of series is not always available.

As regards fractionally integrated ARMA and SARMA models, there are some encouraging results (e.g. Ray, 1993; Sutcliffe, 1994), but the results in Smith and Yadav (1994) suggest that little will be lost by taking first, rather than fractional, differences, while Crato and Ray (1996) say that the use of ARFIMA models for forecasting purposes can only be justified for long time series (meaning at least several hundred observations) which are strongly persistent. Otherwise they say that "simple ARMA models provide competitive forecasts" to FARIMA models.

Many analysts might think that periodic ARMA and ARFIMA models are already complicated enough, but further extensions have been considered. For example, Franses and Ooms (1997) combine the ideas of periodicity and fractional integration by allowing the differencing parameter d to vary with the season to give what is called a *periodic ARFIMA* model. They found evidence from 33 years data on U.K. quarterly inflation that d is significantly larger in the last two quarters of the year. This enabled them to obtain a better fit but, sadly, they found that out-of-sample forecasts were not better than alternative approaches, echoing a major finding regarding (over)complicated models. Many other complicated combinations of models are being looked at, including one paper I saw recently which combined fractional integration with two ideas from non-linear models to give a double threshold, conditionally heteroscedastic ARFIMA model. The problems involved in estimating such a model are formidable and there is no evidence to date that the resulting out-of-sample forecasts are likely to be worth the effort.

Results for non-linear models

Turning now to *non-linear models*, the rewards from using such models can occasionally be worthwhile, but, in the majority of cases, empirical results suggest that non-linear methods are not worth the extra effort as compared with linear methods. Section 3.4 included a brief review of the empirical results on the forecasting accuracy for *threshold* and *bilinear* models, for *models for changing variance* and for *chaos*, and will not be repeated here. Further empirical evidence, in regard to financial time series, is given in the *Journal of Forecasting* (1996, No. 3). The simulation results in Clements and Krolzig (1998) suggest that linear methods are often adequate, even when the data are known to be generated in a non-linear way. Thus it appears that "neither in-sample fit, nor the rejection of the null of linearity in a formal test for non-linearity, guarantee that SETAR models (or non-linear models more generally) will forecast more accurately than linear AR models" (Clements and Smith, 1997, p. 463). Gooijer and Kumar (1992) also suggest that the evidence for non-linear models giving better out-of-sample forecasts is at best mixed.

Given that non-linear models are generally more difficult to fit than linear models (see Section 3.4), and that it can be difficult to compute forecasts more than one step ahead (see Section 4.2.4), it is understandable that many research studies focus attention on the theoretical and descriptive aspects of non-linear models, rather than on forecasting. Much work has been done on theoretical properties of different non-linear models, on estimation and testing, and on modelling issues. For example, it certainly seems to be the case that non-linear models are better suited for modelling long financial series rather than shorter series, such as sales data and many economic series. Indeed, in many cases, it seems that the benefits of non-linear modelling (and also of multivariate modelling!) arise from the thought that needs to be given to modelling issues rather than from improved forecasts.

Results for neural nets

One particular class of non-linear models are *neural networks* (NNs). We consider their forecasting ability separately, in more depth, because of their novelty, the special interest they have generated and the impressive claims that have been made on their behalf. NNs are unlike other non-linear time-series models, in that there is usually no attempt to model the innovations, and much work has been carried out by computer scientists, rather than by statisticians.

NN modelling is nonparametric in character and it is sometimes claimed that the whole process can be automated on a computer "so that people with little knowledge of either forecasting or neural nets can prepare reasonable forecasts in a short space of time" (Hoptroff, 1993). Put another way, this means that NNs can 'learn' the behaviour of a system and produce a good fit and good forecasts in a *black-box* way. While this could be seen as an advantage, black-box modelling is also potentially dangerous. Black boxes can sometimes give silly results and NN models obtained like this are no exception. In my experience (Faraway and Chatfield, 1998), a good NN model for time-series data must be selected by combining traditional modelling skills with knowledge of time-series analysis and of the particular problems involved in fitting NN models. Many questions need to be tackled when fitting a neural net, such as what explanatory variables to include, what structure (architecture) to choose (e.g. how many neurons and how many hidden layers?), how the net should be fitted, and so on.

Earlier reviews of the empirical evidence regarding NNs are given by Faraway and Chatfield (1998), Hill et al. (1994; 1996) and Zhang et al. (1998). One important empirical study was the so-called *Santa Fe* competition (Weigend and Gershenfeld, 1994) where six series were analysed. The series were very long compared with most time series that need to be forecasted (e.g. 34,000 observations) and five were clearly non-linear when their time plots were inspected. There was only one economic series. The organizers kept holdout samples for three of the series. Little contextual information was provided for participants and so I decided not to take part myself. Participants chose their own method of forecasting. As well as using neural nets, some participants chose methods unfamiliar to most statisticians such as the *visual matching of segments* or *multivariate adaptive regression splines*. The results showed that the better NN forecasts did comparatively well, but that some of the worst forecasts were also produced by NNs when applied in 'black-box' mode without using some sort of initial data analysis before trying to fit an appropriate NN model. The results also showed that there are "unprecedented opportunities to go astray". For the one economic series on exchange rates, there was a "crucial difference between training set and test set performance" and "out-of-sample predictions are on average worse than chance". In other words, better forecasts could have been obtained with the Random Walk!

Other empirical evidence suggests that while neural nets may give better forecasts for long series with clear non-linear properties (e.g. the sunspots data), the evidence for shorter series is indecisive. Simulations show linear methods do better than neural nets for data generated by a linear mechanism, as one would intuitively expect. For many economic series, a 'No change' forecast is still found to be better than using a non-linear model or relying on 'experts', while the results of Church and Curram (1996) suggest that neural nets give forecasts which are comparable to (but no better than) those from econometric models, and that "whichever approach is adopted, it is the skill of choosing the menu of explanatory variables which determines the success of the final results". Callen et al. (1996) found that linear methods gave better out-of-sample forecasts than NNs for some seasonal financial series containing 296 observations even though the series appeared non-linear. While it may be too early to make a definitive assessment of the forecasting ability of NNs, it is clear that they are not the universal panacea that some advocates have suggested. Although they can be valuable for long series with clear non-linear characteristics, it appears that the analyst needs several hundred, and preferably several thousand, observations to be able to fit an NN with confidence. Even then, the resulting model is usually hard to interpret, and there is plenty of scope for going badly wrong during the modelling process. For series shorter than about 200 observations, NNs cannot be recommended.

Concluding remarks

It is increasingly difficult to keep up with the wealth of empirical evidence appearing in the literature. Fortunately (or unfortunately depending on your viewpoint), empirical results with real data rarely show anything surprising. An inappropriate method (e.g. using a non-seasonal method on seasonal data) will typically give poor forecasts, as expected, but it is often found that there is little to choose between the forecasting accuracy of several sensible methods. Indeed, whenever I see claims that one method is much better than all alternatives, I get suspicious. Close examination of the results may show that the 'winner' has some unfair advantage. For example, the winning forecasts may not be genuinely out-of-sample (see Section 1.4) or alternative methods may have been implemented poorly or silly alternatives have been chosen. The possibilities for 'cheating' in some way are endless. Sometimes the analyst manages to hide this by not presenting enough background information to make the study replicable. This is also to be deplored.

A related point is that empirical results tend to be reported when they are 'exciting', but quietly dropped otherwise.[2] For example, several studies have been reported that show neural nets can do as well as, or better than, existing forecasting methods. However, Chatfield (1995c) reported at least one study where poor results with neural nets were suppressed, and this may be a more general phenomenon than realized. Sadly, we will never know what has not been reported. Thus it is essential to be aware of potential biases in reported results and treat some results with caution.

6.5 Choosing an appropriate forecasting method

This section gives general advice on how to choose an appropriate forecasting method for a given forecasting problem from the rather bewildering choice available. It updates earlier advice in Chatfield (1988a;

---

[2] This is similar to the well-known publication bias whereby significant statistical results are easier to publish than non-significant ones.

1996a, Section 5.4). Empirical and theoretical results both suggest that there is no single 'best' forecasting method, but rather that the choice of method depends on the context, and in particular on the objectives, the type of data being analysed, the expertise available, the underlying model thought to be appropriate, the number of series which have to be forecasted and so on. The answer to the question "Which is best?" also depends on what is meant by 'best' – see Sections 6.2 and 6.3.

There is a wide variety of problems requiring different treatment, and the sensible forecaster will naturally wish to be discriminating in order to try to solve a given problem (e.g. to get more accurate forecasts). For example, Tashman and Kruk (1996) investigated various protocols for choosing a forecasting method and demonstrated, as might be expected, that substantial gains in accuracy can sometimes be made by choosing a method that is appropriate to the given situation. As a trivial example, some forecasting competitions have applied simple exponential smoothing to all series in the given study regardless of whether they exhibited trend or not. This hardly seems fair to simple exponential smoothing as it does not pretend to be able to cope with trend.

Unfortunately, it is not easy to give simple advice on choosing a forecasting method. A good illustration of the difficulty in making general recommendations is provided by Collopy and Armstrong (1992) who list no less than 99 rules to help make the most appropriate choice from among four univariate methods for annual data. Even so, it is not clear whether an expert system based on such a complicated set of rules will lead to better forecasts. Past empirical work is also only of limited help. The results of forecasting competitions apply mainly to the use of automatic univariate methods on large groups of disparate series and need not generally apply to problems involving a small number of datasets or a homogeneous group of series such as sales of similar items in the same company (Fildes, 1992). Likewise, the results of case studies, using a small number of series, need not apply to data in other application areas. Thus the analyst cannot avoid the difficult task of choosing a method using the context and background knowledge, a preliminary examination of the data, and perhaps a comparative evaluation of a short list of methods.

The first key question in choosing a forecasting method is whether to use (i) an automatic univariate method, (ii) a non-automatic univariate method, or (iii) a multivariate method. We consider these three cases in turn.

A simple *automatic univariate method*, such as exponential smoothing, should be used when there are a large number of series to forecast, when the analyst's skill is limited, or such an approach is otherwise judged to be appropriate for the context and the client's needs and level of understanding. It can also be helpful as a norm for comparison with more complicated forecasts, and as a preliminary forecast to be adjusted subjectively. The first general comment is that the analyst still has to be careful to choose a sensible method, even if it is to be simple and automatic.

Thus it would be ridiculous to apply simple exponential smoothing to data showing trend and seasonality. The empirical results, reviewed in Chatfield (1988a), suggest that there is little overall difference in accuracy between several methods (although one or two more obscure methods should be discarded). Thus it seems sensible to choose a method which is simple, easily interpreted, appropriate for the type of data and for which software is readily available. My personal recommendation is to use an appropriate form of exponential smoothing (e.g. Holt-Winters for seasonal data) but there are several good alternatives.

Some analysts have proposed procedures for using Box-Jenkins in automatic mode, but empirical results suggest that alternative simpler methods have comparable accuracy. In any case, the main virtue of Box-Jenkins ARIMA modelling is to provide the flexibility for finding an appropriate model for a given set of data. Thus, using Box-Jenkins in automatic mode seems to go against the whole rationale of the method.

Suppose now that a *non-automatic* approach is indicated, perhaps because the number of series is small, the analyst's skill is more advanced, and there is contextual information which needs to be taken into account. Good forecasters will use their skill and knowledge to interact with their client, incorporate background knowledge, have a careful 'look' at the data, and generally use any relevant information to build a suitable model so as to construct good forecasts. There are then two possibilities, namely to use a non-automatic type of univariate method, or a multivariate method. We consider the latter approach first.

Multivariate methods are worth considering when appropriate expertise is available, and when there is external information about relationships between variables which needs to be incorporated into a model. This is the case if suitable explanatory variables have been identified and measured, especially when one or more of them is a *leading indicator*. Multivariate models are particularly useful if they are to be used, not only for making forecasts, but also to help gain understanding of the system being modelled. In other words, the model is to be used for description and interpretation, as well as forecasting.

We have seen that a multivariate model can usually be found to give a better *fit* to a given set of data than the 'best' univariate model. However, out-of-sample forecasts from multivariate models are not necessarily more accurate than those from univariate models either in theory or practice, for the reasons listed in Section 5.1. It is worth examining these reasons in turn to see if, how, and when they can be overcome.

(i) More parameters to estimate means more opportunities for parameter uncertainty to affect multivariate forecasts. This suggests aiming for a parsimonious model. It is a common mistake to try to include too many explanatory variables.

(ii) More variables to measure means more opportunities for the data to be affected by errors and outliers. This reminds us that it is essential to carry out careful data cleaning as part of the initial data analysis of both the response and explanatory variables.

(iii) Many time series are of an *observational* nature, rather than coming from a *designed* experiment. This may make them unsuitable for fitting some multivariate models. Open-loop data are easier to fit and likely to give more reliable results, especially if a transfer function model is fitted.

(iv) One way to avoid having to forecast exogenous variables is to use explanatory variables which are leading indicators, whenever possible.

(v) The more complicated the model, the more liable it is to mis-specification and to change. Like (i) above, this suggests aiming for as simple a model as possible. Of course, multivariate models are inherently more complicated than univariate models, and one reason why univariate methods sometimes outperform multivariate methods is that univariate models are more robust to model misspecification and to changes in the model than multivariate models.

As summarized more fully in Section 6.4, multivariate out-of-sample forecasts are more accurate than univariate extrapolations in about half the case studies I have seen, but the reverse is true in the remainder. This contest remains unresolved. Even if multivariate forecasts are better, the question arises as to whether they are worth the extra effort entailed. This can be difficult to assess. However, despite these cautionary remarks, there will be of course be occasions when it does seem promising to try a multivariate approach. The question then arises as to which family of models should be tried.

It is difficult to make general remarks about the choice of a multivariate model as it depends so much on the context, the skill and knowledge of the analyst and the software available. Despite all the dangers outlined in Section 5.2.1, my impression is that *multiple regression* models are still the most commonly used multivariate model, no doubt because of simplicity and familiarity. This is a pity, and explains why I devoted so much space to this family of models in Chapter 5, even though I do not usually recommend their use in time-series forecasting. I really do urge the reader to make sure that he or she understands the reasons why multiple regression models are usually inappropriate for time-series data.

Once the analyst understands that most time series exhibit autocorrelation *within* series, and that this may affect the cross-correlations *between* series, the reader will be well on the way to realizing that multiple regression models (with independent errors) are usually inappropriate. Then more complicated models, such as transfer function and VAR models, which *do* allow for both auto- and cross-correlation, are a more sensible choice. Recent research on multivariate time-series models has focussed on VAR models, and, in particular, on cointegration. With developing methodology and more widely available software, VAR models are becoming much more of a practical possibility, particularly as they make intuitive sense (as do univariate AR models). Before fitting a VAR model to economic data, it is usually wise to check if cointegration is present. If it is, then the cointegration relationship(s) should be incorporated into the model. Before fitting a VAR model to experimental data, it is advisable to ask if the explanatory variables are such as to produce an open-loop system. If they are then a transfer function model should be considered, as this special type of VAR model has been particularly effective in modelling data where the 'outputs' do not affect the 'inputs'.

On many occasions, a 'half-way' house is needed between the complexities of a multivariate approach and the over-simplification of automatic univariate extrapolation. Then a non-automatic *univariate* method should be considered.

The first general point to make is that there is often no clear distinction between an automatic and a non-automatic univariate approach. Few methods are completely automatic in the sense that the analyst must decide, for example, whether to allow for seasonality in the method. Likewise, the analyst can choose to put a lot of effort into making a particular non-automatic forecast, or not very much. For example, the analyst can devote a lot of effort looking for outliers, or just have a quick look at a time plot. He or she can devote a lot of time to assessing the type of trend (e.g. by carrying out a test for unit roots), or choose between no trend, a damped trend and a linear trend by just looking at the time plot. Thus there is really a near-continuous scale measuring the degree of effort employed in applying a particular univariate method, from fully automatic to a high degree of subjective judgement and intervention. What is clear is that fully automatic forecasts can easily be improved with a little skill and effort (e.g. Chatfield, 1978).

One possibility is to use a Box-Jenkins ARIMA-modelling approach. This has advantages and disadvantages. The Box-Jenkins approach has been very influential in the development of time-series modelling and it can be beneficial to be able to choose from the broad class of ARIMA models. However, the accuracy of the resulting forecasts has not always been as good as hoped for, and some forecasters have found it difficult to interpret sample diagnostic tools so as to get a suitable model. When the variation in the data is dominated by trend and seasonal variation, I would not particularly recommend the Box-Jenkins approach as the effectiveness of the ARIMA-modelling procedure is mainly determined by the choice of differencing procedure for removing trend and seasonal effects. The resulting model will not explicitly describe these features and that can be a drawback. However, when the variation is dominated by short-term correlation effects, then ARIMA modelling can be rewarding.

Note that some writers have suggested that exponential smoothing methods are effectively special cases of the Box-Jenkins approach and should therefore not need to be used. Indeed, Newbold (1997), in a recent review of business forecasting methods, says that there is now little justification for using exponential smoothing for non-seasonal data, given the ease with which ARIMA models can be identified using current computer software. However, I would argue that exponential smoothing does still have an important role to play, because the methods are implemented in a completely different way from the Box-Jenkins approach (see Chatfield and Yar, 1988), and they are much easier to apply when there are a large number of series to forecast. Thus I regard the Box-Jenkins approach as being most suitable when there are a small number of series to forecast, when the variation is not dominated by trend and seasonality, and when the analyst is competent to implement the method.

The above remarks raise the general question as to whether it is better to difference away trend and seasonal effects (as in the Box-Jenkins approach), remove trend and seasonality by some other detrending and seasonal adjustment procedure, or use a method which models these effects explicitly. For example, in structural modelling, the Kalman filter can be used to update the Basic Structural Model so that local estimates of level, trend and seasonality are automatically produced. Such estimates can be invaluable for descriptive purposes as well as for forecasting. On other occasions, the basic goal may be to remove seasonal effects and produce seasonally adjusted figures and forecasts. The latter is a quite different problem. As always the context is the key to deciding which approach to adopt.

Having made these general remarks, Section 6.5.1 presents a more detailed outline strategy for choosing an appropriate non-automatic univariate forecasting method, where such an approach is deemed suitable for the given forecasting problem. This is arguably the commonest type of problem in practice.

6.5.1 A general strategy for making non-automatic univariate forecasts

This subsection presupposes that the forecasting problem is such that the analyst is willing to put in a reasonable amount of effort to get a good forecast, but has decided to use a non-automatic univariate forecasting procedure, rather than a multivariate one. The following remarks describe a general strategy for choosing and implementing such a procedure. The following steps are recommended:

1. Get appropriate background information and carefully define the objectives. The *context* is crucial. The type of forecast required (e.g. Point or interval forecasts? Single-period or cumulative forecasts?) will have a strong influence on the choice of method.

2. Make a *time plot* of the data and inspect it carefully. Look for trend, seasonal variation, outliers and discontinuities. Inspection of the time plot is an essential start to the analysis.

3. Clean the data if necessary by correcting obvious errors, adjusting outliers, and imputing missing observations. The treatment is determined primarily by the context.

4. Try to decide if *seasonal variation* is present, and, if so, whether it is

additive, multiplicative or something else. Consider the possibility of transforming the data so as to make the seasonal effect additive.

5. Try to decide if long-term *trend* is present, and, if so, whether it is a global linear trend, a local linear trend or non-linear.

6. Assess the general structure of the time series and, if possible, allocate to one of the following four types of series.

- (i) *One or more discontinuities is present.* Examples include a sudden change in the level, a sudden change in slope, or a large and obvious change in the seasonal pattern. Such changes are often called *structural breaks* and are further discussed in Section 8.5.5, where Figure 8.1 shows an example time plot with a clear change in seasonal pattern. Any univariate forecasting method will have difficulty in coping with sudden changes like this, and so it may be unwise to try such methods. If the context tells you when and why the structural break has occurred, then it may be possible to apply an ad hoc modification to some extrapolation method, such as applying intervention analysis within a Box-Jenkins approach. Alternatively a multivariate or judgemental approach may be considered.

- (ii) *Trend and seasonality are present.* In many series, the variation is dominated by trend and seasonality. The data in Figure 2.1 are of this type. There is usually little point in applying Box-Jenkins to such data, but rather a simple trend and seasonal method should be used. The Holt-Winters version of exponential smoothing is as good as any. While such a method can be handled automatically, better results can be obtained by a careful choice of the type of seasonality (additive or multiplicative), by carefully choosing starting values for the trend and seasonals and by estimating the three smoothing constants, rather than just guessing them. Structural modelling is an alternative approach that is worth considering.

- (iii) *Short-term correlation is present.* For series showing little trend and seasonality, but instead showing short-term correlation effects, it may be worth trying the Box-Jenkins approach. Many economic indicators are of this form. Examples include the daily economic index plotted earlier in Figure 1.1 and quarterly unemployment in the USA which is plotted later in Figure 7.2.

- (iv) *Exponential growth is present.* Many economic series show exponential growth in the long term. For example, an index, like that plotted in Figure 1.1, would show non-linear growth if plotted annually over, say, 20 years. This type of behaviour is particularly difficult to model and forecast. Three alternative possibilities are to fit a model which includes exponential growth terms, to fit a model to the logarithms of the data (or some other suitable transformation), or to fit a model to percentage changes which are typically more stable – see Example 2.2.

7. Whatever method or model is selected to make forecasts, the analyst needs to carry out post-fitting checks to check the adequacy of the forecast. This is usually done by assessing the properties of the fitted residuals and of the out-of-sample forecast errors.

8. Finally the method or model selected can be used to actually make forecasts. An optional extra is to allow such forecasts to be adjusted subjectively, perhaps because of anticipated changes in other variables.

It is worth stressing that all the above stages are important. None can be skipped. Indeed, as already noted in Section 2.3.7, the treatment of (i) trend and seasonality and (ii) outliers, missing observations and calendar effects can be *more* important than the choice of forecasting method, as is the necessity to clarify objectives carefully.

The alert reader will have noticed that the above discussion has not mentioned two alternative general approaches, namely the use of *non-linear* models and the possibility of *combining* forecasts. As regards the former, my overall impression of their current status, summarized in Sections 3.4 and 6.4.2, is that they constitute a valuable addition to the time-series analyst's toolkit, but that the resulting gains in forecasting accuracy are often modest. As yet, their application to forecasting is likely to be of a specialized nature. As regards the combination of forecasts, it sometimes happens that there are several forecasting methods which appear reasonable for a particular problem. Then it is now well established – see Section 4.3.5 – that more accurate forecasts can often be obtained by taking a weighted average of the individual forecasts rather than choosing a single 'best' method. Unfortunately, this *combination* of methods does not give an interpretable model and there may be difficulties in computing prediction intervals, as opposed to point forecasts. Thus this approach is also restricted to specialized applications.

6.5.2 *Implementation in practice*

This subsection makes some brief remarks on implementing forecasting methods in practice. A key requirement is *good computer software*, but the scene is changing so rapidly that it is difficult to make general remarks on this subject. New packages, and new versions of existing packages, continue to come out at ever-decreasing intervals. A comprehensive listing of forecasting software was given by Rycroft in 1999. In order to keep up to date, the reader is advised to read software reviews such as those in the *International Journal of Forecasting* and the *American Statistician*.

Desirable features of 'good' software include: (i) Flexible facilities for entering and editing data; (ii) Good facilities for exploring data and producing a good clear time plot; (iii) The algorithms should be technically sound and computationally efficient; (iv) The output should be clear and self-explanatory; (v) The software should be easy to use with clear documentation.

Many general statistics packages now include some forecasting capability. For example, MINITAB, GENSTAT, S-PLUS and SPSS will all fit ARIMA models. There are many more specialized forecasting packages, written primarily for PCs. They include Forecast Pro for exponential smoothing, AUTOBOX for Box-Jenkins forecasting, STAMP for structural modelling, MTS for transfer function and VARMA modelling, RATS for time series regression, EViews for VAR and econometric modelling, and TSP for VAR and regression modelling.

One final comment on time-series computing is that it is one area of statistics where different packages may not give exactly the same answer to what is apparently the same question. This may be due to different choices in the way that starting values are treated. For example, when fitting an AR(1) model, conditional least squares estimation treats the first observation as fixed, while full maximum likelihood does not. More generally, the choice of algorithm can make a substantial difference as McCullough (1998) demonstrates when comparing Yule-Walker estimates with maximum likelihood. Fortunately, the resulting differences are usually small, especially for long series. However, this is not always the case as demonstrated by the rather alarming examples in Newbold et al. (1994). Thus the forecaster would be wise to give much more attention to the choice of software, as recommended persuasively by McCullough (2000).

A second general requirement for good forecasting is that the forecaster should work in an environment where good quantitative forecasting is encouraged and appreciated. The recommendations in this book have obviously been concerned with what might be called 'best practice', but it should be recognized that real-life forecasting may not always achieve such standards. Most academic research has concentrated on methodological issues such as developing new forecasting methods or improvements to existing ones. The practitioner may be more interested to ask how forecasting practice (as opposed to theory) has changed in recent years. Are the newer forecasting methods actually being used in a commercial setting? How are they used? Do they work well? How are they evaluated? What software is used? Winklhofer et al. (1996) reviewed a large number of surveys and case studies that have been carried out in order to assess the answers to these and other questions.

It appears that many companies still rely mainly on judgemental methods. There are several reasons for this, including a lack of relevant data (series are often too short to use time-series methods), lack of knowledge and suitable software, and a lack of organizational support. Fortunately, familiarity with quantitative methods is increasing. Sometimes, the objectives and role of forecasting are not as clear as they should be, and this can cause problems when interpreting forecasts. When choosing a forecasting method, commercial users typically rate accuracy as the most important criterion followed by factors such as ease of use, time horizon and the number of items to forecast. However, despite the stress on accuracy, it is not always clear how this is actually measured in practice, and many companies do not have routine procedures in place to assess accuracy, highlight large forecast errors and make any resulting revisions that are indicated. It seems that companies still need further help in choosing an appropriate method for a given situation and that further research is needed to establish the importance (or lack of it) of such variables as the type of data. Winklhofer et al. (1996) go on to suggest several aspects of forecasting practice where further research would be helpful.

Overall, we can say that the implementation of forecast methods in practice calls for good judgement by the forecaster, good support from the parent organization and appropriate access to good computer software.

Example 6.1 A forecasting consultancy. Suppose that a client contacts you for help in producing forecasts for a particular time series. Your response depends in part on whether that person is within your own organisation or outside. Either way, you will need to meet the client face-to-face and ask many questions in order to get sufficient background knowledge. You should not underestimate the time required to do this.

As well as finding out exactly how the forecast will be used and what horizon is required, you will need to find out how many observations are available and how reliable they are, whether any explanatory variables are available and so on. Availability of computer software can also be a major consideration, especially if the client wants to do the actual forecast computation. It really helps this briefing discussion if time plots of any observed time series are available. This can lead to a fruitful discussion about whether trend and seasonality are present, and also to related issues such as the possible presence of calendar effects. It is also much better to discuss the possible presence of outliers with the client rather than to jump in with formal tests of possible outliers. There is often a well-understood explanation for 'funny' values, which the client can explain to you. Similar remarks apply to obvious discontinuities, which may on occasion make formal forecasting unwise.

After all these preliminary matters have been addressed, the forecaster can, at last, turn to the topic that is the subject of this chapter, namely deciding which forecasting method should be used. In my experience, this is often largely determined by the preliminary discussion about context, but, if not, then the guidelines given earlier in this section should be helpful. If no explanatory variables are available, then a univariate forecasting method has to be used. If there are hundreds of series to forecast, then a simple method has to be used. If there is no software available to apply a particular method, then a decision must be taken whether to acquire such software or use an alternative method. Apart from the operational researcher trying to set up an inventory control system, or the econometrician trying to forecast a key economic variable, many forecasting consultancies are likely to fall into the general category of choosing a non-automatic univariate method – see Section 6.5.1. Here the time plot is the key to understanding the properties of the data and choosing an appropriate method. It really is the case that many series show regular trend and seasonality and can therefore be adequately forecast using the Holt-Winters procedure. It is also the case that many economic series show short-term correlation, when ARIMA-modelling may be fruitful, or show exponential growth, when the analysis of percentage changes may help.

Having chosen a method, the duties of the forecaster are still not complete. If the client wants to compute the forecasts, the forecaster needs to check that the client has adequate expertise to do this. Advice on interpreting the results and producing interval forecasts should also be given. If the forecaster is instructed to compute the forecasts, this should be done promptly, the results explained clearly and the assumptions on which any forecast is based should be clearly stated. The most common, but often overlooked, assumption is that the future will continue to behave like the past. If the client knows that this will not be the case (e.g. a competitor is organising a sales campaign; the government is planning to change tax rates), then appropriate modifications may be possible. It can be instructive to give the client more than one set of forecasts based on different, explicitly stated, assumptions.

My final piece of advice is to be ready with your excuses when the actual values arrive to be compared with the forecasts you made! Forecast errors have an uncanny knack of being larger than expected! □

6.6 Summary

We conclude this chapter with the following summary, general comments and closing remarks:

1. There are many different types of forecasting problems requiring different treatments.

2. There is no single 'best' method. The choice depends on such factors as the type of data, the forecasting horizon, and, more generally, on the context. The context may also be crucial in deciding how to implement a method. Techniques may need to be adapted to the given situation, and the analyst must be prepared to improvise.

3. Rather than ask which forecasting method should be used, it may be better to ask what *strategy* should be used, so as to produce sensible, robust and hopefully accurate forecasts in the particular context.

4. A good forecasting strategy starts by clarifying exactly how a forecast will be used, and drawing a careful time plot of the data.

5. Time-series model building is generally a tricky process requiring an iterative/interactive approach. Any fitted model is an approximation and different approximations may be useful in different situations.

6. For univariate forecasting, structural modelling and exponential smoothing methods are strong competitors to ARIMA modelling, especially for data showing large trend and seasonal components.

7. Multivariate models are still much more difficult to fit than univariate ones. Multiple regression remains a treacherous procedure when applied to time-series data.

8. Many observed time series exhibit non-linear characteristics, but non-linear models may not give better out-of-sample forecasts than linear models, perhaps because the latter are more robust to departures from model assumptions. Claims for the superiority of neural network models seem particularly exaggerated.

9. The 'best' model for fitting historical data may not be 'best' for generating out-of-sample forecasts, especially if it has a complex form. Whatever method is used, out-of-sample forecasts are generally not as good as would be expected from within-sample fit – see Chapter 8. The comparative accuracy of different forecasting methods should always be assessed on out-of-sample results.

10. It is always a good idea to end with the so-called *eyeball test*. Plot the forecasts on a time plot of the data and check that they look intuitively reasonable. If not, there is more work to do!

CHAPTER 7

Calculating Interval Forecasts

Earlier chapters (and indeed most of the forecasting literature) have been concerned with methods for calculating *point* forecasts. This chapter turns attention to methods for calculating *interval* forecasts, which we call *prediction intervals*. After describing and reviewing some good and some not-so-good methods for doing this, some general comments are made as to why prediction intervals tend to be too narrow in practice to encompass the required proportion of future observations.

7.1 Introduction

Predictions are often given as point forecasts with no guidance as to their likely accuracy. Even more misleadingly, point forecasts are sometimes given with an unreasonably high number of significant digits implying spurious accuracy. Of course, point forecasts may sometimes appear adequate; for example, a sales manager may request a single 'target' point forecast of demand because he or she is unable or unwilling to cope with the uncertainty posed by an interval forecast. The latter requires a more sophisticated level of understanding. In fact, the sales manager, whether he or she likes it or not, will typically have to face the diametrically opposed risks involved in deciding how much stock to manufacture. Too much will result in high inventory costs, while too little may lead to unsatisfied demand and lost profits. Other forecast users often face a similar quandary and so the provision of interval forecasts *is* helpful to many forecast users even though it may raise potentially awkward questions about the assessment of risk, the preparation of alternative strategies for different possible outcomes, and so on. Thus most forecasters do realize the importance of providing interval forecasts as well as (or instead of) point forecasts so as to:

(i) Assess future uncertainty.

(ii) Enable different strategies to be planned for the range of possible outcomes indicated by the interval forecast.

(iii) Compare forecasts from different methods more thoroughly.

Point (iii) above raises the question as to how forecasts should be compared. For example, is a narrower interval forecast necessarily better? We see later on that the answer is NO. I suggest that the provision of realistic interval forecasts enables the analyst to explore different scenarios based on different assumptions more carefully.

7.1.1 Terminology

The first obvious question is "*What is an interval forecast*"? An interval forecast usually consists of an upper and lower limit between which a future value is expected to lie with a prescribed probability. The limits are sometimes called *forecast limits* (Wei, 1990) or *prediction bounds* (Brockwell and Davis, 1991, p. 182), while the interval is sometimes called a *confidence interval* (e.g. Granger and Newbold, 1986) or a *forecast region* (Hyndman, 1995). This book prefers the more widely used term *prediction interval* (e.g. Abraham and Ledolter, 1983; Bowerman and O'Connell, 1987; Chatfield, 1996a; Harvey, 1989), both because it is more descriptive and because the term 'confidence interval' is usually applied to estimates of (fixed but unknown) parameters. In contrast, a prediction interval (abbreviated P.I.) is an estimate of an (unknown) future value which can be regarded as a random variable at the time the forecast is made. This involves a different sort of probability statement to a confidence interval as discussed, for example, by Hahn and Meeker (1991, Section 2.3) in a non-time-series context.

7.1.2 Some reasons for neglect

Given their importance, it is perhaps surprising and rather regrettable that many companies do not regularly produce P.I.s (e.g. Dalrymple, 1987), and that most economic predictions are still given as a single value (though this is slowly changing). Several reasons can be suggested for this, namely:

(i) The topic has been rather neglected in the statistical literature. Textbooks on time-series analysis and forecasting generally give surprisingly scant attention to interval forecasts and little guidance on how to compute them, except perhaps for regression and ARIMA modelling. There are some relevant papers in statistical and forecasting journals, but they can be demanding, unhelpful or even occasionally misleading or even wrong. This chapter attempts to rectify this problem by developing and updating the literature review given by Chatfield (1993). In particular, the chapter incorporates the research described by Chatfield (1996b) concerning the effects of model uncertainty on forecast accuracy, and hence on the the width of P.I.s. The recent summary in Chatfield (2001) is at a lower mathematical level.

(ii) Another reason why interval forecasts are rarely computed in practice is that there is no generally accepted method of calculating P.I.s except for procedures based on fitting a probability model for which the theoretical variance of forecast errors can be readily evaluated.

(iii) Theoretical P.I.s are difficult or impossible to evaluate for many econometric models, especially multivariate models containing many equations or which depend on non-linear relationships. In any case, when judgemental adjustments are made during the forecasting process (e.g. to forecast exogenous variables or to compensate for anticipated changes in external conditions), it is not clear how the analyst should make corresponding adjustments to interval forecasts.

(iv) A forecasting method is sometimes chosen for a group of series (e.g. in inventory control) in a somewhat ad hoc way, by using domain knowledge and the obvious common properties of the various series (e.g. seasonal or non-seasonal). In this sort of situation, no attempt is made to find a probability model for each individual series. Then it is not clear if P.I.s should be based on the model, if any, for which the method is optimal. In other cases a method may be selected which is not based explicitly, or even implicitly, on a probability model, and it is then unclear how to proceed.

(v) Various 'approximate' procedures for calculating P.I.s have been suggested, but there are justified doubts as to their validity.

(vi) Empirically based methods for calculating P.I.s, which are based on within-sample residuals rather than on theory, are not widely understood, and their properties have been little studied.

(vii) Various methods involving some form of resampling (see Section 7.5.6) have been suggested in recent years but have, as yet, been little used in practice.

(viii) Some software packages do not produce P.I.s at all, partly because of (i) to (iv) above. Others produce them for regression and ARIMA modelling only or use 'approximate' formulae which are invalid.

(ix) Whatever method is used to compute P.I.s, empirical evidence suggests they will tend to be too narrow on average. This is particularly true for methods based on theoretical formulae, though less so for empirically based and resampling methods. The fact that P.I.s are not well calibrated in general is another reason why forecast users may not compute P.I.s, or, if they do, may not trust them.

7.1.3 Computing a simultaneous prediction region

This chapter concentrates attention on computing a P.I. for a single observation at a single time horizon. It does not cover the more difficult problem of calculating a simultaneous prediction region for a set of related future observations, either forecasts for a single variable at different horizons, or forecasts for different variables at the same horizon. Thus if the analyst computes a 95% P.I. for a particular variable for every month over the next year, the overall probability that *at least* one future observation will lie outside its P.I. is (much) greater than 5%, and specialized methods are needed to evaluate such a probability. The reader is referred to Ravishankar et al. (1987, 1991) and Lütkepohl (1993, Section 2.2.3).

7.1.4 Density forecasts and fan charts

One topic, closely related to the computation of P.I.s, is that of finding the entire probability distribution of a future value of interest. This is called *density forecasting*. This activity is easy enough for linear models with normally distributed innovations, when the density forecast is typically taken to be a normal distribution with mean equal to the point forecast and variance equal to that used in computing a prediction interval. Conversely, given the density forecast, it is easy to construct prediction intervals for any desired level of probability.

A rather different situation arises when the forecast error distribution is *not* normal. Then density forecasting is more difficult, but seems likely to become more prevalent as it is more rewarding in the non-normal case. One particular application area for non-normal density forecasting is in forecasting volatility, which is important in risk management. When the forecast error distribution is not normal, it may not be possible to compute density forecasts analytically, and so effort has been expended looking at alternative ways of estimating the different percentiles, or quantiles, of the conditional probability distribution of future values (e.g. Taylor, 1999; Taylor and Bunn, 1999). For example, if the 5th percentile and 95th percentile can be estimated, then a 90% prediction interval lies between these two values. An up-to-date review of density forecasting is given by Tay and Wallis (2000), together with several more specialist articles, in a special issue of the *Journal of Forecasting* (Volume 19, No. 4, July 2000).

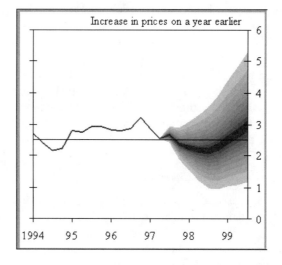

Figure 7.1.   *A fan chart showing prediction intervals for U.K. price inflation (%) up to 2 years ahead from 1997, Q1. The darkest strip covers 10% probability and the lightest covers 90%.*

One interesting development in economic forecasting is the production of what are called *fan charts*. The latter may be thought of as being somewhere between a single prediction interval and a density forecast. The idea is to construct several prediction intervals for different probabilities (e.g. 10%, 30%, 50%, 70%, and 90%), and plot them all on the same graph using different levels of shading to highlight different probabilities. Darker shades are used for central values and lighter shades for outer bands which cover less likely values. If plotted for several steps ahead, the intervals typically 'fan out' as the forecast horizon increases. Fan charts could become a valuable tool for presenting the uncertainty attached to forecasts, especially when the loss function is asymmetric or the forecasts errors are not Gaussian. The origin of fan charts is unclear but they are implicit in Figures 3 and 4 in Thompson and Miller (1986) which give 50% and 90% prediction intervals with shading. Wallis (1999) reviews the work on fan charts carried out by the Bank of England and gives some nice examples, one of which is copied in Figure 7.1. Note how the prediction intervals get wider as the forecast horizon increases.

7.2 Notation

As in earlier chapters, an observed time series, containing N observations, is denoted by $x_1, x_2, \ldots, x_N$, and is regarded as a finite realization of a stochastic process $\{X_t\}$. The point forecast of X_{N+h} made conditional on data up to time N for h steps ahead will be denoted by $\hat{X}_N(h)$ when regarded as a random variable, and by $\hat{x}_N(h)$ when it is a particular value determined by the observed data.

The forecast error conditional on the data up to time N and on the particular forecast which has been made, is given by

$$e_N(h) = X_{N+h} - \hat{x}_N(h) \tag{7.2.1}$$

which is of course a random variable even though $\hat{x}_N(h)$ is not. In future we simply call this the *conditional forecast error*. The observed value of $e_N(h)$, namely $(x_{N+h} - \hat{x}_N(h))$, may later become available. In Section 7.4 we also refer to the unconditional forecast error, namely $E_N(h) = (X_{N+h} - \hat{X}_N(h))$ where both terms on the RHS of the equation are random variables, and so it is arguably more precise to write $e_N(h)$ as

$$e_N(h) = \{X_{N+h} - \hat{x}_N(h) \mid \text{data up to time N}\} \tag{7.2.2}$$

It is important to understand the distinction between the out-of-sample forecast errors, $e_N(h)$, the fitted residuals (or within-sample 'forecast' errors) and the 'error' terms (or innovations) arising in the mathematical representation of the model.

The within-sample one-step-ahead observed 'forecasting' errors, namely $[x_t - \hat{x}_{t-1}(1)]$ for $t = 2, 3, \ldots, N$, are the *residuals* from the fitted model as they are the differences between the observed and fitted values. They will not be the same as the true model innovations because the residuals depend on estimates of the model parameters and perhaps also on estimated starting values. They are also not true forecasting errors when the model parameters (and perhaps even the form of the model) have been determined from all the data up to time N.

As in earlier chapters, we typically use $\{\varepsilon_t\}$ to denote the innovations process so that a model with additive innovations can generally be represented by

$$X_t = \mu_t + \varepsilon_t \qquad (7.2.3)$$

where μ_t describes the predictable part of the model. For example, the predictable part of an AR model is the linear combination of lagged values. Engineers typically refer to μ_t as the *signal* at time t and call $\{\varepsilon_t\}$ the *noise*. The innovations, or noise, are usually assumed to be a sequence of independent normally distributed random variables with zero mean and constant variance σ_ε^2.

If the analyst has found the 'true' model for the data, and if it does not change in the future, then it may seem reasonable to expect the one-step-ahead out-of-sample forecast errors to have similar properties to the residuals, while the variance of h-steps-ahead out-of-sample forecast errors will get larger as the horizon gets longer in a way that may be determined by theory from the within-sample fit. In practice, the out-of-sample forecast errors tend to be larger than expected from within-sample fit, perhaps because the underlying model has changed. We return to this point later.

7.3 The need for different approaches

We have seen that time-series forecasting methods come in a wide variety of forms. They can helpfully be categorized as *univariate* or *multivariate*, as *automatic* or *non-automatic*, and as methods which involve fitting an optimal probability model, and those which do not. The latter distinction is particularly useful when computing P.I.s.

In Chapter 6 we saw that the choice of forecasting method depends on a variety of factors such as the objectives and the type of data. We also saw that the results of previous forecasting competitions apply mainly to the use of automatic methods on large groups of disparate series and need not generally apply when forecasting a single series or a large homogeneous group of series (such as sales of similar items in the same company). Thus the analyst still has the difficult task of choosing a method using background knowledge, a preliminary examination of the data, and perhaps a comparative evaluation of a short list of methods.

Given such a wide range of methods, strategies and contexts, it can be expected that a variety of approaches will be needed to compute P.I.s, and this is indeed the case. In particular, P.I.s for a model-based method can often be computed using theoretical results based on the fitted model. However, for ad hoc methods, P.I.s may need to be based on the empirical properties of the residuals. The various approaches will be introduced in

Section 7.5. All such methods require the analyst to assess the size of prediction errors, and so we first look at ways of doing this.

7.4 Expected mean square prediction error

The usual way of assessing the uncertainty in forecasts of a single variable is to calculate the mean square error for the h-steps-ahead prediction given by $E[e_N(h)^2]$. This quantity is described by the phrase *Prediction Mean Square Error*, which will henceforth be abbreviated to PMSE. If the forecast is unbiased, meaning that $\hat{x}_N(h)$ is the mean of the predictive distribution (i.e. the conditional expectation of X_{N+h} given data up to time N), then $E[e_N(h)] = 0$ and $E[e_N(h)^2] = \text{Var}[e_N(h)]$. Forecasters often assume unbiasedness (explicitly or implicitly) and work with the latter quantity when computing P.I.s – see (7.5.1) below. For many linear models, such as ARMA models, the MMSE point forecast will indeed be unbiased and so much of the literature is devoted to evaluating the PMSE as the *forecast error variance*.

A potential pitfall here is to think that the quantity required to assess forecast uncertainty is the variance of the forecast rather than the variance of the forecast error. In fact, given data up to time N and a particular method or model, the forecast $\hat{x}_N(h)$ will be determined exactly and hence have a conditional variance of zero, whereas X_{N+h} and $e_N(h)$ are random variables, albeit conditioned by the observed data.

At first sight, the evaluation of expressions such as $E[e_N(h)^2]$ or $\text{Var}[e_N(h)]$ may seem to pose no particular problems. In fact, it is not always clear how such expectations should be evaluated and what assumptions should be made. Textbooks rarely consider the problem thoroughly, if at all, though Kendall and Ord (1990, Chapter 8) is a partial exception. There have been a number of technically difficult papers on different aspects of the problem, but they give few numerical illustrations, little qualitative comment and say little or nothing about the construction of P.I.s, which should surely be one of the main objectives.

Consider for simplicity the zero-mean AR(1) process given by

$$X_t = \alpha X_{t-1} + \varepsilon_t \qquad (7.4.1)$$

where $\{\varepsilon_t\}$ are independent $N(0, \sigma_\varepsilon^2)$. Assuming complete knowledge of the model, including the values of α and σ_ε^2, it can be shown (Box et al., 1994, Equation 5.4.16) that

$$E[e_N(h)^2] = \sigma_\varepsilon^2(1 - \alpha^{2h})/(1 - \alpha^2) \qquad (7.4.2)$$

This will be called the 'true-model' PMSE. Formulae for 'true-model' PMSEs can readily be derived for many types of time-series model (see Section 7.5.2).

In practice, even if a model is assumed known, the model parameters will not be known exactly and it will be necessary to replace them with sample estimates when computing both point and interval forecasts. Thus the point forecast $\hat{x}_N(h)$ will be $\hat{\alpha}^h x_N$ rather than $\alpha^h x_N$. When computing interval forecasts, the PMSE or forecast error variance will also need to be estimated. Even assuming that the true model is known a priori (which it will usually not be), there will still be biases in the usual estimate obtained by substituting sample estimates of the model parameters and the residual variance into the true-model PMSE formula (Ansley and Newbold, 1981). Restricting attention to the case $h = 1$ for simplicity, and conditioning on $X_N = x_N$, we consider $e_N(1) = X_{N+1} - \hat{x}_N(1)$. If the model parameters were known, then $\hat{x}_N(1)$ would equal αx_N and $e_N(1)$ would reduce to ε_{N+1}, but as the parameters are *not* known, we find

$$e_N(1) = X_{N+1} - \hat{x}_N(1) = \alpha x_N + \varepsilon_{N+1} - \hat{\alpha} x_N = (\alpha - \hat{\alpha}) x_N + \varepsilon_{N+1} \quad (7.4.3)$$

Finding the expectation of the square of expressions like this is not easy and the rest of this section considers the effect of parameter uncertainty on estimates of the PMSE. We assume throughout that parameter estimates are obtained by a procedure which is asymptotically equivalent to maximum likelihood.

First we look at the expected value of $e_N(1)$ rather than its square. Looking back at (7.4.3), for example, it is clear that if x_N is fixed and $\hat{\alpha}$ is a biased estimator[1] for α, then the expected value of $e_N(1)$ need not be zero (Phillips, 1979). If, however, we average over all possible values of x_N, as well as over ε_N, then it can be shown that the expectation will indeed be zero giving an unbiased forecast. The former operation is *conditional* on x_N while the latter involves the *unconditional* forecast error, namely $E_{N+1}(1) = (X_{N+1} - \hat{X}_N(1))$. It is important to be clear what one is, or is not, conditioning on and it could also be useful to have additional notation to distinguish between forecasts involving true and estimated parameters. There has been much confusion because of a failure to distinguish the different types of situation.

The distinction between conditional and unconditional expectations can also be important when computing PMSE. Many authors (e.g. Yamamoto, 1976, for AR processes; Baillie, 1979, and Reinsel, 1980, for vector AR processes; and Yamamoto, 1981, for vector ARMA processes) have looked at the PMSE by averaging over the distribution of future innovations (e.g. ε_{N+1} in (7.4.3)) and over the distribution of the current observed values, (e.g. x_N in (7.4.3)), to give the *unconditional* PMSE. This approach was also used by Box et al. (1994, Appendix A7.5) to assess the effect of parameter uncertainty on the PMSE when they concluded that correction terms would generally be of order $1/N$. The unconditional PMSE can be useful to assess the 'success' of a forecasting method *on average*. However, if used to compute P.I.s, it effectively assumes that the observations used to estimate the model parameters are independent of those used to construct the forecasts. While this assumption can be justified asymptotically, Phillips

---

[1] Did you realize, for example, that the least-squares estimator for α can have a sizeable bias for short series (Ansley and Newbold, 1980)?

(1979) points out that "it is quite unrealistic in practical situations" and goes on to look at the distribution of the forecast errors for the AR(1) case *conditional* on the final observed value, x_N. The resulting mean square error is called the *conditional* PMSE.

Phillips' results for the conditional PMSE of an AR(1) process have been extended, for example, by Fuller and Hasza (1981) to AR(p) processes and by Ansley and Kohn (1986) to state-space models. As the general ARMA model can be formulated as a state-space model, the latter results also cover ARMA models (and hence AR processes) as a special case. Note that PMSE formulae for regression models are typically of conditional form and *do* allow for parameter uncertainty – see Section 7.5.2.

From a practical point of view, it is important to know if the effect of incorporating parameter uncertainty into PMSEs has a non-trivial effect. Unfortunately, the literature appears to have made little attempt to quantify the effect.

Consider, for example, a K-variable vector AR(p) process, with known mean value. It can be shown that the 'true-model' PMSE at lead time one has to be multiplied by the correction factor $[1 + Kp/N] + o(1/N)$ to give the corresponding unconditional PMSE allowing for parameter uncertainty (e.g. Lütkepohl, 1993, Equation 3.5.13). Thus the more parameters there are, and the shorter the series, the greater will be the correction term as would intuitively be expected. When $N = 50, K = 1$ and $p = 2$, for example, the correction to the square root of PMSE is only 2%. Findley's (1986, Table 3.1) bootstrapping results also suggest the correction term is often small, though results from more complex models suggest it can be somewhat larger. When $N = 30, K = 3$ and $p = 2$, for example, the correction to the square root of PMSE rises to 6%. However, the effect on *probabilities* is much smaller (see Lütkepohl, 1993, Table 3.1). This may be readily demonstrated with the standard normal distribution where 95% of values lie in the range ± 1.96. Suppose the standard deviation is actually 6% larger because the above correction factor has not been used. Then the percentage of values lying inside the range $\pm(1.96 \times 1.06)$ or ± 2.08 is 96.2% rather than 95%. Thus the change in the probability coverage is relatively small. This non-linear relation between corrections to square roots of PMSEs and corrections to the resulting probabilities is worth noting.

In the AR(p) case, the formula for the *conditional* one-step-ahead PMSE will also involve the last p observed values. In particular, when $p = 1$, the correction term involves an expression proportional to x_N^2. Thus if the last observed value is 'large', then the conditional PMSE will be inflated which is also intuitively reasonable.[2]

---

[2] There is a natural analogy here with standard linear regression where the P.I. for a future observation on the response variable at a new value of the explanatory variable, say x^*, relies on an expression for the conditional PMSE which involves a term proportional to $(x^* - \bar{x})^2$ (e.g. Weisberg, 1985, Equation 1.36), where $\bar{x}$ is the average x-value in the sample used to fit the model.

More work needs to be done to assess the size of correction terms in the conditional case for time-series models, but few authors actually use the conditional PMSE in computing P.I.s (except when using regression models), presumably because of the difficulty involved in evaluating the conditional expression and in having to recompute it every time a new observation becomes available.

Overall, the effect of parameter uncertainty seems likely to be of a smaller order of magnitude in general than that due to other sources, notably the effects of model uncertainty and the effect of errors and outliers (see Section 7.7). Thus I agree with Granger and Newbold (1986, p. 158) that "for most general purposes, it should prove adequate to substitute the parameter estimates" into the true-model PMSE. However, for models with a large number of parameters in relation to the length of the observed series, this strategy could lead to a more serious underestimate of the length of P.I.s (Luna, 2000).

7.5 Procedures for calculating P.I.s

This section reviews various approaches for calculating P.I.s.

7.5.1 Introduction

Most P.I.s used in practice are essentially of the following general form. A $100(1 - \alpha)\%$ P.I. for X_{N+h} is given by :

$$\hat{x}_N(h) \pm z_{\alpha/2}\sqrt{\text{Var}[e_N(h)]} \qquad (7.5.1)$$

where $z_{\alpha/2}$ denotes the percentage point of a standard normal distribution with a proportion $\alpha/2$ above it, and an appropriate expression for $\text{Var}[e_N(h)]$ is found from the method or model being used.

As the P.I. in (7.5.1) is symmetric about $\hat{x}_N(h)$, it effectively assumes that the forecast is unbiased with PMSE equal to the forecast error variance (so that $E[e_N(h)^2] = \text{Var}[e_N(h)]$). The formula also assumes that the forecast errors are normally distributed.

When $\text{Var}[e_N(h)]$ has to be estimated (as it usually must), some authors (e.g. Harvey, 1989, p. 32) suggest replacing $z_{\alpha/2}$ in (7.5.1) by the percentage point of a t-distribution with an appropriate number of degrees of freedom, but this makes little difference except for very short series (e.g. less than about 20 observations) where other effects, such as model and parameter uncertainty, are likely to be more serious.

The normality assumption may be true asymptotically but it can be shown that the one-step-ahead conditional forecast error distribution will not in general be normal, even for a linear model with normally distributed innovations, when model parameters have to be estimated from the same data used to compute forecasts. This also applies to h-steps-ahead errors (Phillips, 1979), although the normal approximation does seem to improve as h increases, at least for an AR(1) process. Looking back at (7.4.3), for example, we have already noted that the expected value of the conditional distribution of $e_N(1)$ need not be zero, and Phillips (1979) shows that the conditional distribution will generally be skewed. This is another point to bear in mind when computing P.I.s.

Whether the departure from normality caused by parameter uncertainty is of practical importance seems doubtful. The only guidance offered by Phillips (1979) is that when N is 'small' (how small?), the correction to the normal approximation can be 'substantial', and that the normal approximation becomes less satisfactory for the AR(1) process in (7.4.1) as the parameter α increases in size. However, the correction term is of order $1/N$ and seems likely to be of a smaller order of magnitude in general than that due to model uncertainty, and to the effect of errors and outliers and other departures from normality in the distribution of the innovations. The possibility of departures from normality for reasons other than having to estimate model parameters is considered in Section 7.7.

Rightly or wrongly, (7.5.1) is the formula which is generally used to compute P.I.s, though preferably after checking that the underlying assumptions (e.g. forecast errors are approximately normally distributed) are at least reasonably satisfied. For any given forecasting method, the main problem will then lie with evaluating $\mathrm{Var}[e_N(h)]$.

7.5.2 P.I.s derived from a fitted probability model

If the true model for a given time series is known, then it will usually be possible to derive minimum MSE forecasts, the corresponding PMSE and hence evaluate P.I.s, probably using (7.5.1). The practitioner typically ignores the effect of parameter uncertainty and acts as though the estimated model parameters are the true values. Thus it is the 'true-model' PMSE which is usually substituted into (7.5.1).

Formulae for PMSEs are available for many classes of model. Perhaps the best-known equation is for Box-Jenkins ARIMA forecasting, where the PMSE may be evaluated by writing an ARIMA model in infinite-moving-average form as

$$X_t = \varepsilon_t + \psi_1\varepsilon_{t-1} + \psi_2\varepsilon_{t-2} + \dots \qquad (7.5.2)$$

– see Equation (3.1.3). Then $e_N(h) = [X_{N+h} - \hat{X}_N(h)]$ can be evaluated using Equation (4.2.3) to give $e_N(h) = \varepsilon_{N+h} + \sum_{j=1}^{h-1} \psi_j\varepsilon_{N+h-j}$ so that

$$\mathrm{Var}[e_N(h)] = [1 + \psi_1^2 + \dots + \psi_{h-1}^2]\sigma_\varepsilon^2 \qquad (7.5.3)$$

This is a 'true-model' PMSE and we would of course have to insert estimates of $\{\psi_i\}$ and σ_ε^2 into this equation in order to use it in practice.[3]

---

[3] Note that (7.5.3) involves a finite sum and so will converge even when the sequence of ψ's does not. For a non-stationary ARIMA model, the values of ψ_j for any finite j can be calculated (even though the whole series diverges) by equating coefficients of B^j in the equation $\theta(B) = \phi(B)\psi(B)$ using the notation of Section 3.1.4.

The equations (7.5.1) and (7.5.3) are sometimes the only equations for P.I.s given in textbooks.

Formulae for PMSEs can also be derived for vector ARMA models (e.g. Lütkepohl, 1991, Sections 2.2.3 and 6.5) and for structural state-space models (Harvey, 1989, Equation 4.6.3). PMSE formulae are also available for various regression models (e.g. Weisberg, 1985, Section 1.7; Miller, 1990, Section 6.2; Kendall and Ord, 1990, Equation 12.32), but these formulae *do* typically allow for parameter uncertainty and are conditional in the sense that they depend on the particular values of the explanatory variables from where a prediction is being made. One application of these regression results is to the (global) linear trend model when the explanatory variable is time.

Finally, we mention two classes of model where PMSE formulae may not be available. For some complicated simultaneous equation econometric models, it is not possible to derive $\text{Var}[e_N(h)]$, particularly when some of the equations incorporate non-linear relationships and when judgement is used in producing forecasts (for example, in specifying future values of exogenous variables). Then an empirically based approach must be used as in Sections 7.5.5 and 7.5.6. Equation (7.5.1) is also inappropriate for many non-linear models. In Section 4.2.4, we noted that it can be difficult to evaluate conditional expectations more than one step ahead, and that the forecast error variance need not necessarily increase with lead time. Moreover, the predictive distribution will not in general be normal (or Gaussian) and may, for example, be bimodal. In the latter case a single point forecast could be particularly misleading (see Tong, 1990, Figure 6.1) and a sensible P.I. could even comprise two disjoint intervals. Then the description *forecast region*, suggested in Hyndman (1995), seems more appropriate than the term P.I.

For stochastic volatility and ARCH models, which aim to model the changing variance of a given variable rather than the variable itself, there has been little work on the computation of P.I.s. for the original variable. Most work has concentrated on forecasting the changing variance rather than using this to construct P.I.s for the original variable. The conditional predictive distribution of the original variable will not in general be normal, and so the problem is not an easy one and will not be pursued here.

Thus for non-linear models, there may be little alternative to attempting to evaluate the complete predictive distribution even though this may be computationally demanding. Note that economists also work with what they call non-linear models, though they are different in kind from the mainly univariate models considered in Tong (1990), in that they are typically multivariate and involve non-linear power transformations. Even in the univariate case, the application of a power transformation (e.g. taking logs to make a variable exhibit additive seasonality) may introduce non-linearities (see Section 7.5.8).

Most of the above assumes that the 'true model' for the given time series is known. In practice, the true model is not known and so the fitted model is typically formulated from the data. This is usually done by pre-specifying a broad class of models, such as ARIMA or state-space models, and choosing the most appropriate model from within that class. Unfortunately, the practitioner then typically acts as though the selected model *was* known a priori and is the true model. This inevitably leads to over-optimism in regard to the accuracy of the forecasts as will be explored more fully in Chapter 8. Put another way, this means that P.I.s calculated using PMSEs for the best-fitting model may be poorly calibrated. For example, it is typically found that more than 5% of future observations will lie outside a 95% P.I.

Note that models are customarily fitted by minimizing one-step-ahead 'errors' in some way even when h-steps-ahead forecasts are required. This is valid provided one has specified the correct model, but it is more robust to fit models using a fit statistic appropriate to the forecast requirement, meaning that within-sample h-steps-ahead errors should be minimized. Further comments on this point are given in Section 8.5 – see especially Example 8.6.

7.5.3 *P.I.s derived by assuming that a method is optimal*

A forecasting method is sometimes selected without applying any formal model identification procedure (although one should certainly choose a method which does or does not cope with trend and seasonality as appropriate). The question then arises as to whether P.I.s should be calculated by some empirical procedure (see Sections 7.5.5 – 7.5.6) or by assuming that the method is optimal in some sense.

Consider exponential smoothing (ES), for example. This well-known forecasting procedure can be used for series showing no obvious trend or seasonality without trying to identify the underlying model. Now ES is known to be optimal for an ARIMA$(0, 1, 1)$ model or for the random walk plus noise model, and both these models lead to the PMSE formula (Box et al., 1994, p. 153; Harrison, 1967)

$$\text{Var}[e_N(h)] = [1 + (h - 1)\alpha^2]\sigma_e^2 \qquad (7.5.4)$$

where α denotes the smoothing parameter and $\sigma_e^2 = \text{Var}[e_n(1)]$ denotes the variance of the one-step-ahead forecast errors. Should this formula then be used in conjunction with (7.5.1) for ES even though a model has not been formally identified? My answer would be that it is reasonable (or at least not unreasonable) to use (7.5.4) provided that the observed one-step-ahead forecast errors show no obvious autocorrelation and provided that there are no other obvious features of the data (e.g. trend) which need to be modelled.

There are, however, some P.I. formulae for ES which I would argue should be disregarded. For example, Abraham and Ledolter (1983) follow Brown (1963) in deriving formulae for various smoothing methods based on a general regression model. In particular, it can be shown that ES arises from the model

$$X_t = \beta + \varepsilon_t \qquad (7.5.5)$$

when β is estimated by discounted least squares and expressed as an updating formula. Then P.I.s can be found for model (7.5.5) assuming that β is constant. Although these formulae can take account of the sampling variability in $\hat{\beta}$, they have the unlikely feature that they are of constant width as the lead time increases (e.g. Abraham and Ledolter, 1983, Equation 3.60; Bowerman and O'Connell, 1987, p. 266). Intuitively this is not sensible. It arises because β is assumed constant in (7.5.5). But if this were true, then ordinary least squares should be used to estimate it. The use of discounted least squares suggests that β is thought to be changing. If indeed β follows a random walk, then we are back to the random walk plus noise model for which (7.5.4) is indeed appropriate. More generally, the many formulae given in the literature based on General Exponential Smoothing derived by applying discounted least squares to global models such as (7.5.5) should be disregarded, since ordinary least squares is optimal for a global model (Abraham and Ledolter, 1983, p. 126). As a related example, Mckenzie (1986) derives the variance of the forecast error for the Holt-Winters method with additive seasonality by employing a deterministic trend-and-seasonal model for which Holt-Winters is not optimal. These results should also be disregarded.

Some forecasting methods are not based, even implicitly, on a probability model. What can be done then? Suppose we assume that the method is optimal in the sense that the one-step-ahead errors are uncorrelated (this can easily be checked by looking at the correlogram of the one-step-ahead errors; if there is correlation, then there is more structure in the data which it should be possible to capture so as to improve the forecasts). From the updating equations, it may be possible to express $e_N(h)$ in terms of the intervening one-step-ahead errors, namely $e_N(1), e_{N+1}(1), \ldots, e_{N+h-1}(1)$. If we assume that the one-step-ahead errors are not only uncorrelated but also have equal variance, then it should be possible to evaluate $\mathrm{Var}[e_N(h)]$ in terms of $\mathrm{Var}[e_N(1)]$. It may also be possible to examine the effects of alternative assumptions about $\mathrm{Var}[e_N(1)]$.

Yar and Chatfield (1990) and Chatfield and Yar (1991) have applied this approach to the Holt-Winters method with additive and multiplicative seasonality, respectively. In the additive case it is encouraging to find that the results turn out to be equivalent to those resulting from the seasonal ARIMA model for which additive Holt-Winters is optimal (although this model is so complicated that it would never be identified in practice). The results in the multiplicative case are of particular interest because there is no ARIMA model for which the method is optimal. Chatfield and Yar (1991) show that, assuming one-step-ahead forecast errors are uncorrelated, $\mathrm{Var}[e_N(h)]$ does not necessarily increase monotonically with h. Rather P.I.s tend to be wider near a seasonal peak as would intuitively be expected. More self-consistent results are obtained if the one-step-ahead error variance is assumed to be proportional to the seasonal effect rather than constant. The phenomenon of getting wider P.I.s near a seasonal peak is not captured by most alternative approaches (except perhaps by using a variance-stabilizing transformation). The lack of monotonicity of $\mathrm{Var}[e_N(h)]$ with h is typical of behaviour resulting from non-linear models (see comments in Section 7.5.2 and Tong, 1990, Chapter 6) and arises because multiplicative Holt-Winters is a non-linear method in that forecasts are not a linear combination of past observations. Note that Ord et al. (1997) have investigated a dynamic nonlinear state-space model for multiplicative Holt-Winters which allows model-based P.I.s to be calculated.

7.5.4 P.I.s based on 'approximate' formulae

For some forecasting methods and models, theoretical P.I. formulae are not available. As one alternative, a variety of approximate formulae have been suggested, either for forecasting methods in general or for specific methods. Because of their simplicity, they have sometimes been used even when better alternatives *are* available. This is most unfortunate given that the approximations turn out to have poor accuracy in many cases. Some readers may think the proposed approximations are too silly to warrant serious discussion, but they *are* being used and do need to be explicitly repudiated.

(i) One general 'approximate' formula for the PMSE is that

$$\mathrm{Var}[e_N(h)] = h\sigma_e^2 \qquad (7.5.6)$$

where $\sigma_e^2 = \mathrm{Var}[e_N(1)]$ denotes the variance of the one-step-ahead forecast errors. This formula is then substituted into (7.5.1) to give P.I.s. Equation (7.5.6) is given by Makridakis et al. (1987, Equation 1), Lefrancois (1989, Equation 1) and verbally by Makridakis and Winkler (1989, p. 336). It is stated to depend on an unchanging model having independent normal errors with zero mean and constant variance. In fact, these assumptions are not enough and (7.5.6) is true only for a random walk model (Koehler, 1990). For other methods and models it can be seriously in error (e.g. see Table 7.1 in Section 7.6 and Yar and Chatfield, 1990) and should not be used.

(ii) Equation (7.5.6) may have arisen from confusion with a similar-looking approximation for the error in a cumulative forecast (see Lefrancois, 1990, and the reply by Chatfield and Koehler, 1991). Let

$$e_N^C(h) = e_N(1) + \ldots + e_N(h) \qquad (7.5.7)$$

= cumulative sum of forecast errors at time N over next h periods.

Brown (1963, p 239) suggests verbally that

$$\mathrm{Var}[e_N^C(h)] = \sum_{i=1}^{h} \mathrm{Var}[e_N(i)] \qquad (7.5.8)$$

and by assuming that $\mathrm{Var}[e_N(i)]$ is a constant (!??) it is but a short step to the approximation (e.g. Johnston and Harrison, 1986, p. 304) that

$$\text{Var}[e_N^C(h)] = h\sigma_e^2 \qquad (7.5.9)$$

However, (7.5.8) ignores the correlations between errors in forecast made from the same time origin and Brown's statement has been the source of much confusion. There is no theoretical justification for (7.5.9), which, like (7.5.6), can give very inadequate results.

(iii) Brown (1967, p. 144) also proposes an alternative approximation for the variance of the cumulative error, namely that

$$\text{Var}[e_N^C(h)] = (0.659 + 0.341h)^2\sigma_e^2 \qquad (7.5.10)$$

This approximation, like (7.5.6) and (7.5.9), cannot possibly be accurate for all methods and models and in some cases is seriously inadequate.

Makridakis et al. (1987, Equation 2) have cited Brown's formula but applied it, not to the cumulative error, but to a single forecast error, so that they effectively take

$$\text{Var}[e_N(h)] = (0.659 + 0.341h)^2\sigma_e^2 \qquad (7.5.11)$$

This appears to be a simple error from misreading Brown's book and is another example of the confusion between single-period and cumulative forecasts (Chatfield and Koehler, 1991). Equation (7.5.11) should therefore not be used.

(iv) Only one example of an approximation aimed at a specific method will be given here. Bowerman and O'Connell (1987, Section 6.4) give approximate formulae for P.I.s for the Holt-Winters method. The formulae are rather complicated and depend on the maximum of the three smoothing parameters. As such they appear to be producing conservative limits in some way but the exact reasoning behind these formulae is unclear. They are not compatible with the exact results given by Yar and Chatfield (1990) and Chatfield and Yar (1991). (Note that Bowerman and O'Connell's (1987) formulae look unfamiliar because they effectively estimate σ_e as 1.25 times the mean absolute one-step-ahead forecasting error over the fit period which is a standard alternative to the use of root mean square error.)

7.5.5 Empirically based P.I.s

When theoretical formulae are not available, or there are doubts about the validity of the 'true' model anyway, the reader should not use the so-called 'approximate' formulae, but should use a more computationally intensive approach based either on using the properties of the *observed* distribution of 'errors', as described in this subsection, or based on simulation or resampling methods (see Section 7.5.6).

(i) One simple empirically based type of procedure (e.g. Gilchrist, 1976, p. 242) involves applying the forecasting method to all the past data, finding the within-sample 'forecast' errors at 1, 2, 3, ... steps ahead from all available time origins, and then finding the variance of these errors at each lead time over the period of fit. Let $s_{e,h}$ denote the standard deviation of the h-steps-ahead errors. Then, assuming normality, an approximate empirical $100(1 - \alpha)\%$ P.I. for X_{N+h} is given by $\hat{x}_N(h) \pm z_{\alpha/2}s_{e,h}$, where we replace $\sqrt{\mathrm{Var}[e_N(h)]}$ by $s_{e,h}$ in (7.5.1). If N is small, $z_{\alpha/2}$ is sometimes replaced by the corresponding percentage point of the t-distribution with ν degrees of freedom, where $s_{e,h}$ is based on ν degrees of freedom. However, there is no real theoretical justification for this. A reasonably long series is needed in order to get reliable values for $s_{e,h}$. Even so, it may be wise to smooth the values over neighbouring values of h, for example, to try to make them increase monotonically with h. The approach often seems to work reasonably well, and gives results comparable to theoretical formulae when the latter are available (Yar and Chatfield, 1990; Bowerman and Koehler, 1989). However, the values of $s_{e,h}$ can be unreliable for small N and large h, and are based on in-sample residuals rather than on out-of-sample forecast errors. There is evidence that the latter tend to have a larger variance – see Section 7.7. Thus P.I.s calculated in this way may be calibrated poorly (as may P.I.s calculated in other ways).

(ii) An earlier related method (Williams and Goodman, 1971) involves splitting the past data into two parts, fitting the method or model to the first part and make predictions of the second part. The resulting 'errors' are much more like true forecast errors than those in (i) above, especially for long series where model parameters can be estimated with high precision. The model is then refitted with one additional observation in the first part and one less in the second part; and so on. For some monthly data on numbers of business telephone lines in service, Williams and Goodman found that the distribution of forecast errors tended to approximate a gamma distribution rather than a normal distribution. P.I.s were constructed using the percentage points of the empirical distribution, thereby avoiding any distributional assumptions. Promising results were obtained. However, although the approach is attractive in principle, it seems to have been little used in practice, presumably because of the heavy computational demands. However, the latter problem has not prevented developments as in Section 7.5.6, and it may be that the Williams-Goodman method was ahead of its time and should now be reassessed.

7.5.6 Simulation and resampling methods

This type of approach is even more computationally intensive than that described in Section 7.5.5, but is increasingly used for the construction of P.I.s (and many other problems). The approach can be used when theoretical formulae are not available, for short series when only asymptotic results are available and when there are doubts about model assumptions.

Given a probability time-series model, it is possible to *simulate* both past and future behaviour by generating an appropriate series of random innovations and hence constructing a sequence of possible past and future values. This process can be repeated many times, leading to a large set of possible sequences, sometimes called pseudo-data. From such a set it is possible to evaluate P.I.s at different horizons by finding the interval within which the required percentage of future values lie. This approach was tried out by Ord et al. (1997) for a particular nonlinear state-space model underlying the multiplicative Holt-Winters method and found, for both simulated and real data, that a simulation method of computing P.I.s gave better coverage than alternative approximation methods. The use of simulation is sometimes called a *Monte Carlo* approach. It generally assumes that the model has been identified correctly.

Instead of sampling the innovations from some assumed parametric distribution (usually normal), an alternative is to sample from the empirical distribution of past fitted 'errors'. This is called *resampling* or *bootstrapping*. The procedure effectively approximates the theoretical distribution of innovations by the empirical distribution of the observed residuals. Thus it is a distribution-free approach. As for simulation, the idea is to use the knowledge about the primary structure of the model so as to generate a sequence of possible future values and find a P.I. containing the appropriate percentage of future values by inspection. It may also be possible to extend the use of resampling to forecasting methods which are not based on a proper probability method, but rely instead on a set of recursive equations involving observed and forecast values.

For the reader who has not come across bootstrapping before, the term is usually used to describe the process of taking a random sample of size n from a sample of independent observations of the same size n, where observations are selected *with replacement*. This means that some observations will occur twice in the bootstrap sample while others will not occur at all. In a time-series context, this type of sampling would make no sense because successive observations are not independent but are correlated through time. This explains why time-series data are usually bootstrapped by resampling the fitted errors (which the analyst will hope are at least approximately independent) rather than the actual observations. However, the reader should be aware that it is generally more difficult to resample correlated data, such as time series, rather than resample independent observations. Moreover, the effect of resampling the fitted errors makes the procedure much more dependent on the choice of model which has been fitted. Some model-free methods have been proposed (e.g. Carlstein, 1992) but I think they are harder to understand and technically more difficult.

The literature in this area, particularly on bootstrapping, is growing rapidly. Veall (1989) reviews the use of computationally intensive methods for complex econometric models, where they are particularly useful, and gives many references. He suggests that "most applied econometric exercises should include bootstrapping or some other form of simulation as a check". Bootstrapping can, of course, be used for other aspects of time-series-analysis such as evaluating the standard errors of estimates of model parameters (e.g. Freedman and Peters, 1984a, 1984b), where the normality assumption may be less critical than in the evaluation of P.I.s (Veall, 1989, Section 3.2). However, it can be a mistake to concentrate on departures from the secondary model assumptions (e.g. normal errors), when departures from the primary assumptions (or specification error) can be much more serious.

One classic simulation study reviewed by Veall (1989) is that of Fair (1980) who showed how to assess four sources of uncertainty in forecasts from econometric models namely (i) the model innovations; (ii) having estimates of model parameters rather than true values; (iii) having forecasts of exogenous variables rather than true values; and (iv) misspecification of the model. Fair sampled from a multivariate normal distribution for two example models, one a large (97 equations!) model and the other a much simpler autoregressive model. Fair pointed out that assessing the uncertainty due to misspecification is the most difficult and costly part of model assessment and his approach rested on some rather restrictive assumptions.

We now concentrate on references not covered by Veall (1989). Early work on the use of resampling for calculating P.I.s (e.g. Butler and Rothman, 1980) was for regression models and not really concerned with time-series forecasting. Freedman and Peters (1984b, Section 6) give one example of forecasting no less than 24 years ahead and show how to compute what they call "standard errors of forecasts". Peters and Freedman (1985) show how to use bootstrapping to get standard errors for multi-step-ahead forecasts for the complex 10-equation model of Freedman and Peters (1984a). They show that the results are more reliable for short series than those given by the so-called delta method (Schmidt, 1977). Note that exogenous variables are forecast "by some process external to the equation". Bianchi et al. (1987) discuss various simulation and resampling methods as applied to a large macro model of the French economy involving over 20 equations.

There have been several papers which are specifically concerned with AR models. Findley (1986) shows how to compute bootstrap estimates of both the unconditional and conditional PMSE for an $AR(p)$ process. As forecasts for an $AR(p)$ model depend on the last p observed values, Findley says that it is the error associated with conditional predictions from sample paths through these last p observations which is usually of most interest, but went on to say that satisfactory methods for obtaining such sample paths were not then available. Latterly Stine (1987) and Thombs and Schucany (1990) have both shown how to overcome this problem for AR processes by using the *backward* representation of the series conditional on the last p observations. For example, in the zero-mean AR(1) case, fix y_N equal to the latest observation, x_N, and generate backward sample paths from

200 CALCULATING INTERVAL FORECASTS

(c.f. (7.4.1))

$$y_t = \hat{\alpha} y_{t+1} + \hat{\varepsilon}_t \tag{7.5.12}$$

for $t = (N - 1), (N - 2), \ldots$ where $\hat{\alpha}$ is the least-squares estimate of α from the (original) observed series and $\hat{\varepsilon}_t$ are independent samples from the empirical distribution of the observed backward residuals. Each sample path can then be used to re-estimate the parameter α in (7.4.1), after which conditional bootstrap replicates of the future can be constructed using the bootstrap estimate of α and further random drawings from the empirical distribution of forward residuals. Full details are given by Thombs and Schucany (1990). Stine (1987) looked at unconditional P.I.s for $AR(p)$ processes as well as conditional P.I.s, and carried out various simulations. He showed that bootstrap P.I.s compare favourably with normal-based P.I.s, particularly when the innovations are not normal as would intuitively be expected. Thombs and Schucany (1990) also simulated various AR(1) and AR(2) models with innovations which are normal, exponential or a mixture of two normals. They also concluded that bootstrap P.I.s are a useful non-parametric alternative to the usual Box-Jenkins intervals. Masarotto (1990) also looked at bootstrap P.I.s for AR models but appears to only have considered the unconditional (and less interesting?) case. Most authors have assumed that the order of the AR process is known, but Masarotto does explicitly take into account the possibility that the order, as well as the model parameters, are generally unknown. Masarotto presented simulation results for AR(1), AR(3) and AR(5) processes with innovations which are normal or from the centered extreme value distribution. One feature of interest is that innovations were sampled not only from a parametric distribution and from the empirical distribution of residuals, but also from a smoothed version of the latter using a kernel density estimate. There seems to be little difference in the results from the last two types of distribution, so that the extra computation needed to smooth the empirical residual distribution was not worthwhile in this case. McCullough (1994) applied bootstrapping to the calculation of 'forecast intervals' for $AR(p)$ models and found "substantial differences between Box-Jenkins, naive bootstrapping and bias-correction bootstrapping" though the differences look less than substantial to me. McCullough (1996) later applied bootstrapping to the derivation of P.I.s for multiple regression models when future-period values of exogenous variables are not known with certainty, and claimed successful results. Grigoletto (1998) has proposed a new method for finding P.I.s for AR models which takes account of the additional uncertainty induced by estimating the order of the model as well as the model parameters.

The book by Hjorth (1994) includes discussion of methods for bootstrapping time-series, together with applications to road safety, forecasting the stock market and meteorology. Hjorth's work is also referred to in Chapter 8 because of his investigation of the effects of model selection uncertainty.

Monte Carlo simulation has been used by Pflaumer (1988) to calculate P.I.s for population projections by letting fertility and net immigration rates vary as random variables with specified distributions. This is arguably superior to the 'alternative scenarios' approach where high, medium and low assumptions are made about different components leading to a range of possible population trajectories. However, no probabilities can be attached to the latter and a single P.I. may well be easier to interpret. An alternative approach for a large (non-linear) econometric model, which involves stochastic perturbation of input variables, is described by Corker et al. (1986). Finally, we mention that Thompson and Miller (1986) also used simulation and their paper is referred to in Sections 7.5.7 and 7.8.

As a closing comment, it should also be said that bootstrapping is not always successful. For example, the problem of finding P.I.s for growth curve models is difficult because of the non-linear nature of the model and because it is not always clear how to sensibly specify the model error structure. Meade and Islam (1995, p. 427) tried bootstrapping growth curves but found that it gave P.I.s which were "far too narrow" and so preferred an alternative procedure, called the 'explicit density approach' which gave more plausible (asymmetric) intervals.

7.5.7 The Bayesian approach

If the analyst is willing to consider adopting a Bayesian approach, then the following brief summary of Bayesian possibilities should be read. Although the author of this book is not a Bayesian (meaning someone who insists on using the Bayesian approach for every problem), a fair assessment of the approach will be attempted. Many statisticians try to avoid labels like 'Bayesian' and 'frequentist' and adopt whatever approach seems sensible for a particular problem.

In principle, given a suitable model, the Bayesian approach will allow the complete probability distribution for a future value to be computed. From this distribution it should be possible to derive interval forecasts, either by a decision-theoretic approach along the lines of Winkler (1972), or (more usually) by calculating symmetric intervals, using the Bayesian version of (7.5.1), when the predictive distribution is normal. If the predictive distribution has some other symmetric distribution (e.g. Student t), then (7.5.1) can readily be adapted by inserting appropriate percentiles. Unfortunately, the multi-period ahead predictive density does not have a convenient closed form for many forecasting models, and so Bayesian statisticians may need to use some sort of approximation when interval forecasts are required (e.g. Thompson and Miller, 1986, Section 3). Alternatively, it is possible to simulate the predictive distribution, rather than try to obtain or approximate its analytic form. Thompson and Miller (1986) compare the resulting percentiles of the simulated predictive distribution for an AR(2) process with the Box-Jenkins P.I.s based on (7.5.1) and (7.5.3). As the latter do not allow for parameter uncertainty, the Bayesian intervals are naturally somewhat wider (although they still do not include all the ensuing observed values – see Section 7.8).

West and Harrison (1997) and Pole et al. (1994) provide a comprehensive treatment of *Bayesian forecasting* based on a general class of models called *dynamic linear models*. The latter are much more amenable to the Bayesian approach than ARIMA models. Unfortunately, the two books say little explicitly about interval forecasts although they are computed in several examples. It seems to be assumed implicitly that the mean and variance of the forecast distribution are substituted into (7.5.1) together with normal or t-percentiles as appropriate. The results are variously called 'probability limits', 'prediction intervals', 'symmetric intervals', 'intervals for the one-step-ahead forecast' and 'forecast intervals' by West and Harrison (1997), and 'uncertainty limits' by Pole et al. (1994). If the error terms in the observation and system equations are assumed to be normal with *known* variances and if (conjugate) normal priors are assumed, then the distribution of forecast errors will also be normal. However, the error variances will generally be unknown (as will the parameters in the corresponding ARIMA model) and if they are allowed to evolve as a normal process, then a Student t-distribution will result. All in all, there are a lot of assumptions but the identification process seems (to me) to lack the cohesive strength of the Box-Jenkins approach. There will, of course, be certain situations where a 'dynamic' or local model is indicated, but then the non-Bayesian may prefer the conceptual approach of Harvey's (1989) structural modelling.

A Bayesian approach does seem very natural when the analyst decides to rely, not on a single model (which is likely to be a good approximation at best), but on a *mixture* of models. The idea for this follows on from the well-known result that *combining* forecasts from different methods and models (e.g. by averaging) generally gives more accurate point forecasts on average (e.g. Clemen, 1989) than any of the constituent point forecasts. Unfortunately, there is no simple analytic way of computing P.I.s for a combined forecast of this type, although it should be possible to use some sort of resampling method or an empirical approach (Taylor and Bunn, 1999). An alternative to combining forecasts in a more-or-less ad hoc way is to use Bayesian methods to find a sensible set of models which appear plausible for a given set of data, and then to average over these models in an appropriate way. While similar in spirit, this approach has a stronger theoretical foundation than simply combining forecasts, and is called *Bayesian model averaging*. It will be reviewed in Section 8.5.3, where the emphasis will be on its use for overcoming the effects of model uncertainty. Here we simply present a brief summary of the main example in Draper (1995, Sections 2.2 and 6.1) which is particularly concerned with getting a sensible P.I.

Draper's example is concerned with forecasting oil prices, and is particularly instructive in demonstrating that conditioning on a single model or scenario can seriously underestimate the uncertainty in forecasts, because the effects of model uncertainty are not allowed for. The problem was to forecast the 1986 price per barrel of crude oil from data up to 1980. Ten different econometric models were considered together with twelve different scenarios embodying a variety of assumptions about the input to the models, such as low demand elasticity or a drop in OPEC production. Thus 120 different point forecasts could be made, which could then be averaged across a particular scenario, across a particular model, or across all scenarios and models. The point forecasts averaged across a particular scenario ranged from $29 to $121. Although this range is (staggeringly?) high, the P.I.s for an individual scenario did not reflect this level of uncertainty. For example, the 90% P.I. for the reference scenario was from $27 to $51. In fact, the actual price in 1986 turned out to be $11 per barrel!! This was a long way outside all the different P.I.s that were computed. The prime reason for this was that the P.I.s did not reflect uncertainty about the model. Draper went on to use a Bayesian model averaging approach which gave a 90% P.I. of ($20, $92) when taking account of both model and scenario uncertainty. This is much wider, and therefore better (meaning more realistic), than the P.I.s found for individual models and individual scenarios. However, it was still not wide enough! Further comments on this example follow in Section 8.4.1.

7.5.8 P.I.s for transformed variables

Whichever approach is used to calculate P.I.s, the possibility of working with a transformed variable needs to be considered (e.g. Granger and Newbold, 1986, Section 10.5; West and Harrison, 1997, Section 10.6). It may be sensible for a variety of reasons to work, not with the observed variable X_t, but with some non-linear transformation of it, say $Y_t = g(X_t)$, where g may, for example, be the logarithmic transformation or the more general Box-Cox transformation which is defined in Section 2.3.3. A transformation may be taken in order to stabilize the variance, to make the seasonal effect additive, to make the data more normally distributed or because the transformed variable makes more sense from a practical point of view. For example, logs of economic variables are often taken when percentage growth is of prime interest.

P.I.s may be calculated for Y_{N+h} in an appropriate way, but the literature says very little about transforming the P.I.s back to get P.I.s for the original observed variable if this is desired. (Collins, 1991 is an exception but he considers regression models.) It is well known that the 'naive' point forecast of X_{N+h}, namely $g^{-1}[\hat{y}_N(h)]$, is generally not unbiased, essentially because the expected value $E[g^{-1}(Y)]$ is not generally equal to $g^{-1}[E(Y)]$. If the predictive distribution of Y_{N+h} is symmetric with mean $\hat{y}_N(h)$, then $g^{-1}[\hat{y}_N(h)]$ will be the *median*, rather than the mean, of the predictive distribution of X_{N+h}. Fortunately, the situation regarding P.I.s for a retransformed variable is more straightforward. A 'naive' P.I. can be constructed by retransforming the upper and lower values of the

P.I. for Y_{N+h} using the operator g^{-1}. If the P.I. for Y_{N+h} has a prescribed probability, say $(1-\alpha)$, then it is immediately clear that the retransformed P.I. for X_{N+h} should have the *same* prescribed probability (e.g. Harvey, 1990, Equation 4.6.7) apart from any additional uncertainty introduced by trying to identify the transformation, g. Fortunately, the results of Collins (1991) suggest that uncertainty about the Box-Cox transformation parameter may be relatively unimportant. As an aside, it is worth noting that Collins' results indicate that model parameter estimates are likely to be correlated with the transformation parameter when the latter has to be estimated. It is also worth noting that Collins uses the description 'plug-in density' to describe forecasts arising from replacing unknown parameters with their estimated values, and the term 'deterministic prediction' to describe forecasts made by applying the inverse Box-Cox transformation to forecasts of Y_{N+h} which *do* allow for parameter uncertainty, though I do not understand why this is termed 'deterministic'.

If the P.I. for Y_{N+h} is based on a normality assumption, and hence is symmetric, then the transformed P.I. for X_{N+h} will be asymmetric (which is often intuitively sensible). The width of the transformed P.I. will depend on the level as well as the lead time and the variability. It may be possible to derive a symmetric P.I. for X_{N+h} (Collins (1991) refers to this as a "mean-squared error analysis"), but this is not sensible unless X_t is thought to be normally distributed – in which case a transformation will probably not be sensible anyway. In regard to this point, I note that some forecasting software I have seen does retransform both the point forecast and its standard error in order to compute symmetric P.I.s using (7.5.1). This is definitely not recommended!

Although point forecasts may be relatively little affected by whether or not a transformation is used, the P.I.s will typically be affected rather more by the use of transformation, particularly in becoming asymmetric. My own preference, stemming from my early experiences with transformations and the problems they can cause (Chatfield and Prothero, 1973), is to avoid transformations wherever possible except where the transformed variable is of interest in its own right (e.g. taking logarithms to analyse percentage increases) or is clearly indicated by theoretical considerations (e.g. Poisson data has non-constant variance). Thus I agree with West and Harrison (1997, p. 353) that "uncritical use of transformations for reasons of convenience should be guarded against" and that generally "it is preferable to model the series on the original data scale".

7.5.9 Judgemental P.I.s.

Judgement may be used in time-series forecasting, not only to produce point forecasts, but also to produce P.I.s. Empirical evidence (e.g. Armstrong, 1985, pp. 138-145; O'Connor and Lawrence, 1989, 1992; Webby and O'Connor, 1996) suggests that the P.I.s will generally be too narrow, indicating over-confidence in the forecasts. In a special issue on 'Probability

Judgemental Forecasting' in the *International Journal of Forecasting* (1996, No. 1), Wright et al. (1996) summarize the evidence on the accuracy and calibration of judgemental prediction intervals as being "not very encouraging". Judgemental forecasting, even when applied to time series, is rather outside the scope of this book and so will not be pursued here.

7.6 A comparative assessment

The choice of procedure for calculating P.I.s in a particular situation depends on various factors. The most important of these is the choice of forecasting method which depends in turn on such factors as the objectives and type of data. However, there will sometimes be a choice of methods for computing P.I.s and this section attempts a general assessment of how to decide which one to choose. Note that Yar and Chatfield (1990) made a more restricted comparison of the different approaches which could be used for additive Holt-Winters.

Theoretical P.I. formulae based on (7.5.1), with the PMSE determined by a fitted probability model, are easy to implement, but do assume that the fitted model has been correctly identified, not only in regard to the primary assumptions (e.g. X_t is a linear combination of lagged variables as in an AR process), but also in regard to the secondary 'error' assumptions (e.g. the innovations are independent $N(0, \sigma^2)$). Allowance can be made for parameter uncertainty but the formulae are complex and generally give corrections of order $1/N$ and so are rarely used. A more serious problem is that the model may be misspecified or may change in the forecast period. Nevertheless, the formulae are widely used, although they are not available for some complex and/or non-linear models.

Formulae which simply assume that a given forecasting method is optimal (see Section 7.5.3) are also widely used because of their simplicity, but it is then important to check that the method really is a sensible one (see Section 7.7). Formulae based on a model for which the method is *not* optimal should be disregarded.

Empirically based and resampling methods are always available and require fewer assumptions, but can be much more computationally demanding (especially when resampling is used). Nevertheless, they have much potential promise, particularly when theoretical formulae are not available or there are doubts about the error assumptions. However, it is important to remember that resampling does still usually depend on the *primary* assumptions made regarding the selected model. For example, Thombs and Schucany (1990) compute bootstrap P.I.s under the assumption that an $AR(p)$ model really does fit the data even though they make no assumptions about the 'error' component of the model. We also remind the reader that resampling time-series data, which are naturally ordered in time, is usually more difficult than resampling other types of data where order is not important. It can also be difficult to tell whether the results of a resampling exercise are intuitively reasonable, given the

h	"Approximate" [RW] (7.5.6)	(7.5.11)	AR(1) $\phi = 0.5$	MA(1) $\theta = 0.5$	ARIMA (1,1,0) $\phi = 0.5$	[ARIMA (0,1,1)] ES $\alpha = 0.3$	$\alpha = 0.7$	[ARIMA (0,2,2)] Holt Linear $\gamma = 0.1$ $\alpha = 0.3$	$\alpha = 0.9$
2	2	1.8	1.250	1.25	3.2	1.09	1.5	1.1	2.0
4	4	4.1	1.328	1.25	9.8	1.27	2.5	1.4	4.5
8	8	11.5	1.333	1.25	25.3	1.63	4.4	2.2	12.3
12	12	22.6	1.333	1.25	41.3	1.99	6.4	3.6	24.7

Table 7.1 *Values of Var $[e_N(h)]/$ Var $[e_N(1)]$ for various equations, models and methods. Notes: The "approximate" formulae in Equations (7.5.6) and (7.5.11) are h and $(0.659 + 0.341h)^2$, respectively; formula (7.5.6) gives the same ratios as the random walk (RW) or ARIMA(0,1,0) model; the AR(1), MA(1) and ARIMA(1,1,0) models use the notation of Section 3.1; the smoothing parameters for exponential smoothing (ES) and Holt's linear smoothing use the notation of Section 4.3; ES is optimal for an ARIMA(0,1,1) model and Holt's Linear method for an ARIMA(0,2,2) model.*

absence of a theoretical basis to check results. When a researcher tries several different methods of resampling a time series, gets rather different results and chooses the one which seems intuitively most reasonable (as I have seen take place) the outcome is less than convincing. Generally speaking, this is an area where further research is particularly required.

It is hard to envisage any situation where the approximate formulae should be used. They have no theoretical basis and cannot possibly capture the varied properties of P.I.s resulting from different methods. Table 7.1 shows the ratio of Var$[e_N(h)]$ to Var$[e_N(1)]$ for the two approximate formulae in (7.5.6) and (7.5.11) as well as the theoretical results for various methods and models. The disparate relationship with h is evident. For stationary models, such as an AR(1) or MA(1) model, the width of P.I.s will increase rather slowly with h to a finite upper bound (or even be constant for the MA(1) model), whereas for non-stationary models the width of P.I.s will increase without bound. This applies to the random walk and ARIMA(1,1,0) models, to exponential smoothing (optimal for an ARIMA(0,1,1) model) and to Holt's linear trend method (optimal for an ARIMA(0,2,2) model). The inadequacy of (7.5.6) and (7.5.11) is clear.

7.7 Why are P.I.s too narrow?

Wallis (1974) noted many years ago that it is a "common experience for models to have worse error variances than they should when used in forecasting outside the period of fit". Subsequent empirical studies have generally borne this out by showing that out-of-sample forecast errors tend to be larger than model-fitting residuals so that P.I.s tend to be too narrow on average. In other words, more than 5% of future observations will typically fall outside the 95% P.I.s on average. The evidence is reviewed by Fildes and Makridakis (1995) and Chatfield (1996b) and includes, for example, results reported by Makridakis and Winkler (1989), by Newbold and Granger (1974, p. 161) for Box-Jenkins models, by Williams and Goodman (1971) for regression models, by Gardner (1988) in respect of empirical formulae, and by Makridakis et al. (1987) for Box-Jenkins models.[4] More generally the "in-sample fit of a model may be a poor guide to ex-ante forecast performance" (Clements and Hendry, 1998b) and out-of-sample forecast accuracy tends to be worse than would be expected from within-sample fit. Further remarks on this important phenomenon are made in Chapter 8 (especially Example 8.2).

There are various possible reasons why P.I.s are too narrow, not all of which need apply in any particular situation. They include:

(i) Model parameters have to be estimated;

(ii) For multivariate forecasts, exogenous variables may have to be forecasted;

(iii) The innovations may not be normally distributed, but could, for example, be asymmetric or heavy-tailed. The latter effect could be due to occasional outliers. There may also be errors in the data which will contaminate the apparent 'error' distribution.

(iv) Unconditional, rather than conditional, P.I.s are typically calculated;

(v) The wrong model may be identified;

(vi) The underlying model may change, either during the period of fit or in the future.

Problem (i) can often be dealt with by using PMSE formulae incorporating correction terms for parameter uncertainty, though the corrections are typically of order $1/N$ and smaller than those due to other sources of uncertainty.

Problem (ii) goes some way to explaining why multivariate forecasts need not be as accurate as univariate forecasts, contrary to many people's intuition (e.g. Ashley, 1988). P.I.s for multivariate models can be found which take account of the need to forecast other endogenous variables, and also exogenous variables when the latter are forecast using a separate model (Lütkepohl, 1993, Section 10.5.1). However, when future values of the exogenous variables are assumed known, or are guessed (e.g. assumed to grow at a constant inflation rate), then the P.I.s will not take account of this additional uncertainty.

Problem (iii) is very common. For non-linear models, it is generally the case that the innovations are known not to be normal anyway (e.g.

---

[4] The latter study is flawed in its use of 'approximate' formulae (see Section 7.6), but the results for one-step-ahead forecasts and for Box-Jenkins should still be relevant.

Hyndman, 1995) and so the following remarks are aimed primarily at linear models. The analyst will typically be concerned about two types of departure from normality, namely (i) asymmetry and (ii) heavy tails. Some empirical evidence (e.g. Makridakis and Winkler, 1989) suggests neither is a serious problem when averaged over many series, but there is evidence of asymmetry in other studies (e.g. Williams and Goodman, 1971; Makridakis et al., 1987) and asymmetry can sometimes be expected from theoretical considerations. Many measured variables are inherently non-negative (i.e. have a natural zero) and show steady growth (in that the average *percentage* change is approximately constant). Then it is typically found that the residuals from a fitted model are skewed to the right. This applies particularly to annual economic data. The residuals can often be made (more) symmetric by taking logs and by taking account of explanatory variables (Armstrong and Collopy, 2001). If a model is formulated for the logs, and then used to compute point and interval forecasts for future values of the logged variable, then these will need to be transformed back to the original units in order to give forecasts of what is really required – see Section 7.5.8. The resulting P.I.s will generally be asymmetric.

One alternative to transforming the data is to modify (7.5.1) by changing the Gaussian percentage point, $z_{\alpha/2}$, to a more appropriate value for describing the 'error' distribution, when standardized to have unit variance. This can be done parametrically by utilizing a probability distribution with heavier tails than the normal. Possible distributions include the t-distribution and the stable distribution. The latter can be generalized to cope with asymmetry (Lambert and Lindsey, 1999). A non-parametric approach is also possible, and there is extensive literature on the treatment of heavy tails (e.g. Resnick, 1997). The simplest type of approach, already mentioned briefly in Section 7.5.6, is to find appropriate upper and lower percentage points of the empirical standardized error distribution for one-step-ahead forecasts for the raw data, say $z^\star_{1-\alpha/2}$ and $z^\star_{\alpha/2}$, and use them to modify 7.5.1 in an obvious way. The revised $100(1-\alpha)\%$ P.I. for X_{N+h} is given by $(\hat{x}_N(h) + z^\star_{1-\alpha/2}\sqrt{\mathrm{Var}[e_N(h)]},\ \hat{x}_N(h) + z^\star_{\alpha/2}\sqrt{\mathrm{Var}[e_N(h)]})$, bearing in mind that lower percentage points, such as $z^\star_{1-\alpha/2}$, will be negative.

Note that, if outliers arise as a result of large errors in the data, then this will affect the perceived distribution of innovations even if the true underlying distribution really is normal. More generally, the presence of outliers and errors can have an effect out of all proportion to the number of such observations, both on model identification, and on the resulting point forecasts and P.I.s (e.g. see Ledolter, 1989). The effect is particularly marked when the outlier is near the forecast origin, but fortunately "the impact of outliers that occur well before the forecast origin is usually discounted rapidly" (Ledolter, 1989, p. 233).

Problem (iv) reminds us that, even when the innovations are normal, the conditional prediction errors need not be normal, but will typically be asymmetric and have a larger variance than those calculated in the usual unconditional way. Forecasters generally use unconditional forecasts, because of their simplicity. However, more attention should be given to the possibility of using the more reliable conditional forecasts, especially for ARCH-type models, where volatility can change. Prediction intervals need to be wider when behaviour is more volatile, and then the difference between conditional and unconditional interval forecasts can be substantial. Christofferson (1998) describes a framework for evaluating conditional interval forecasts.

Problems (v) and (vi), relating to model uncertainty, are discussed more fully in Chapter 8. Problem (v) may arise for various reasons. In particular, it is always tempting to overfit the data with more and more complicated models in order to improve the fit but empirical evidence suggests that more complicated models, which give a better fit, do not necessarily give better forecasts. Indeed, it is strange that we admit model uncertainty by searching for the best-fitting model, but then ignore such uncertainty by making forecasts as if the fitted model is known to be true. As we see in Chapter 8, one reason why model uncertainty is often ignored in practice is that there is no general theoretical way of taking it into account. It is, however, worth noting that the use of bootstrapping may be able to take account of the possibility of identifying the wrong model from within a given class (e.g. Masarotto, 1990).

Problem (v) should, of course, be circumvented whenever possible by carrying out appropriate diagnostic checks. For example, when fitting ARIMA models, model checking is an integral part of the identification process (Box et al., 1994, Chapter 8). Even when using a forecasting method which does not depend explicitly on a probability model, checks should still be made on the (possibly implicit) assumptions. In particular, checks should be made on the one-step-ahead fitted errors to see if they (i) are uncorrelated (ii) have constant variance. It may be sufficient to calculate the first-order autocorrelation coefficient and the autocorrelation at the seasonal lag (if there is a seasonal effect). If the values are significantly different from zero (i.e. exceed about $2/\sqrt{N}$ in modulus), then this suggests that the optimal method or model is not being used and there is more structure to find. To check constant variance, it is a good idea to compare the residual variances in the first and second halves of the data and also to compare periods near a seasonal peak with periods near a seasonal trough. It often seems to be the case that the residual variance increases with the mean level and is higher near a seasonal peak. These features need to be explicitly dealt with (e.g. see Chatfield and Yar, 1991) or alternatively the data could be transformed so as to stabilize the variance (see Section 7.5.8).

Problem (vi) may arise for all sorts of reasons which can often be classified into one of two possible states, namely that the underlying structure is evolving slowly through time or that there is a sudden shift or turning point. As regards the first possibility, there is plenty of empirical evidence that the economies of many countries evolve through time, often rather slowly, and that this effect manifests itself in fixed specification linear models by finding that estimates of model parameters evolve through time (e.g. Swanson and White, 1997). Two economic examples of the second possibility are the sudden changes which occurred in many economic variables as a result of the 1973 oil crisis and the 1990 Gulf war. Change points can have a particularly devastating effect on forecast accuracy and so the prediction of change points is a topic of continuing interest even though it is notoriously difficult to do (see Makridakis, 1988). There is therefore much to be said for using a forecasting method which adapts quickly to real change but which is robust to occasional 'blips' or outliers. Unfortunately, it is not easy to devise a method which gets the balance 'right' as between reacting to a permanent change but not reacting to a temporary fluctuation.

For all the above reasons, post-sample forecast errors tend to be larger than model-fitting errors, as already noted in Section 7.5.5 and earlier in this section. This explains why it is essential that different forecasting models and methods are compared on the basis of out-of-sample forecasts rather than on measures of fit. Because P.I.s tend to be too narrow, Gardner (1988) suggested modifying (7.5.1) to

$$\hat{x}_N(h) \pm \sqrt{\mathrm{Var}[e_N(h)]}/\sqrt{\alpha} \qquad (7.7.1)$$

where the constant $1/\sqrt{\alpha}$ (which replaces $z_{\alpha/2}$) is selected using an argument based on Chebychev's inequality. However, Bowerman and Koehler (1989) point out that this may give very wide P.I.s in some cases which are of little practical use. In any case they may be unnecessarily wide for reasonably stable series where the usual normal values will be adequate. On the other hand, when there is a substantial change in the forecast period (e.g. a change in trend), then the Chebychev P.I.s may still not be wide enough.

The use of 50%, rather than 95%, P.I.s has been suggested (e.g. Granger, 1996) as being more robust to outliers and to departures from model assumptions and this will also help overcome the problem of computing P.I.s. only to find they are embarrassingly wide. However, using a probability of 50% will mean that a future value is equally likely to lie inside or outside the interval. This seems undesirable.

My own preference is generally to use (7.5.1), incorporating the normality assumption, rather than (7.7.1), but, as a compromise, to use 90% (or perhaps 80%) intervals rather than 95%. I also recommend stating explicitly that this assumes (i) the future is like the past, with all the dangers that entails; (ii) the errors are approximately normal. If an observation does fall outside a computed P.I., it is not necessarily a disaster, but can indeed be enlightening. It may, for example, indicate a change in the underlying model. The possible diagnostic use of P.I.s in this sort of way deserves attention and may perhaps usefully be added to other diagnostic tools for detecting trouble (e.g. Gardner, 1983).

But whatever checks are made and whatever precautions are taken, it is still impossible to be certain that one has fitted the correct model or to rule out the possibility of structural change in the present or future, and problems (v) and (vi) above are, in my view, the most important reasons why P.I.s are too narrow. Section 7.8 gives an instructive example where two plausible models give substantially different P.I.s for the same data.

7.8 An example

This example is designed, not to compare different approaches to computing P.I.s (as reviewed in Section 7.6), but to illustrate the overriding importance of careful model identification. The data shown in Figure 7.2 were analysed by Thompson and Miller (1986) to compare Box-Jenkins P.I.s with a Bayesian approach which simulates the predictive distribution. The fourth quarter in 1979 was taken as the base month and forecasts were computed up to 12 steps (3 years) ahead.

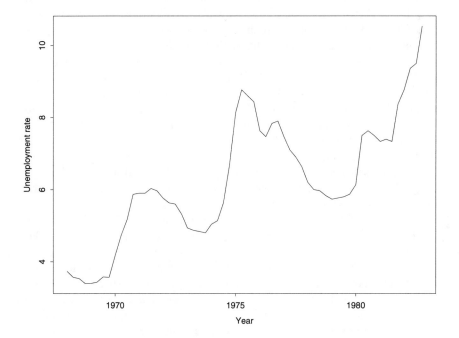

Figure 7.2. *Time plot of U.S. quarterly unemployment rate (%) 1968–1982 (seasonally adjusted). The model-fitting period ends in 1979, Q4.*

The point forecasts computed by Thompson and Miller (1986) were generally poor because of the large (unforeseeable?) increase in unemployment. The Bayesian P.I.s were somewhat wider than the Box-Jenkins P.I.s because the Bayesian method allows for parameter uncertainty, and this is clearly advantageous here. However, the Bayesian P.I.s still fail to include the actual values for 10, 11 and 12 steps ahead.

Thompson and Miller go on to hypothesize a shift in level which does produce better forecasts. But could this hypothesis reasonably have been made in 1979? If it could, then perhaps a multivariate model, including any known leading variable(s), should have been constructed. (Indeed, some readers may think such a model is intrinsically more sensible anyway than a univariate model, but that is not the main point of this example).

Thompson and Miller (1986) identified an AR(2) model with a non-zero mean. However, Figure 5.1 suggests to me that the series is non-stationary, rather than stationary. This view is reinforced by the sample autocorrelation function (ac.f.) of the first 12 years data that decreases slowly and is still positive up to lag 9. The ac.f. of the first differences decreases much faster, with a significant value at lag one (0.65), while the value at lag two (0.28) is nearly significant (c.f. $\pm 2/\sqrt{48} \simeq 0.29$). Subsequent values are small with the value at lag three equal to 0.08. This suggests an AR(1) model for the first differences or, equivalently, an ARIMA(1, 1, 0) model for the original data. The latter model was fitted, and the resulting sum of squared residuals is a little higher (4.08 rather than 3.64 for the AR(2) model), but diagnostic checks on the residuals suggest the model is adequate (the modified Box-Pierce χ^2 statistic of 12.3 on 11 D.F. is actually better than the 12.2 on 10 D.F. for the AR(2) model).

The point forecasts from the ARIMA(1, 1, 0) model are perhaps a little better than those from the AR(2) model (e.g. 6.00 versus 5.69 for 12 steps ahead), but it is the P.I.s which concern us here. At 12 steps ahead, for example, the 95% P.I. for the ARIMA(1, 1, 0) model is (0.86, 11.14) while that for the AR(2) model is (2.66, 8.73). Thus the P.I.s from the ARIMA(1, 1, 0) model are much wider and *do* include the actual values. Wide P.I.s are sometimes seen as indicating 'failure', either to fit the right model or to get a usable interval, but here the wider P.I.s are arguably more realistic in allowing for high uncertainty.

The crucial difference between an AR(2) and an ARIMA(1, 1, 0) process is that the first is stationary, but the second is non-stationary. For stationary processes, $\mathrm{Var}[e_N(h)]$ tends to the variance of the process as $h \to \infty$. In other words, P.I.s will tend to a constant finite width as the lead time increases. But for non-stationary processes, there is no upper bound to the width of P.I.s. This is noted in passing by Box et al. (1994, p. 159) but the result is perhaps not as widely appreciated as it should be. The higher the degree of differencing needed to make a series stationary, the larger in general will be the rate at which the width of the resulting P.I.s. will diverge. Thus for an ARIMA(0, 1, 1) process, $\mathrm{Var}[e_N(h)]$ is of order h, while for an ARIMA(0, 2, 2) process, $\mathrm{Var}[e_N(h)]$ is of order h^3 [Box et al., 1994, Equations (5.4.7) and (5.4.14)].

There is a similar dichotomy for multivariate time-series models. For stationary series, the PMSE for each series tends to its variance, while that for non-stationary series increases without bound (e.g. Lütkepohl, 1993, p. 377), although it is relevant to note that for non-stationary series which are 'tied together' or cointegrated, a multi-step forecast which satisfies the cointegrating relationship will have a *finite* limiting PMSE (Engle and Yoo, 1987).

Returning to the example, the wider P.I.s for the non-stationary ARIMA$(1,1,0)$ process seem to capture the observed uncertainty better than the narrower stationary alternative, given that a univariate model is to be fitted. The difference in the P.I. widths is much larger than that resulting from parameter uncertainty, for example, and confirms the over-riding importance of model identification, particularly in regard to deciding whether the data are stationary or not.

7.9 Summary and recommendations

The computation of interval forecasts is of vital importance in planning and deserves more attention. A variety of approaches for computing P.I.s have been described and compared. The main findings and recommendations can be summarized as follows:

(i) A theoretically satisfying way of computing prediction intervals (P.I.s) is to formulate a model which provides a reasonable approximation for the process generating a given series, derive the resulting prediction mean square error (PMSE), and then use (7.5.1). Formulae for the PMSE are available for various classes of models including regression, ARIMA and structural models. A correction term to allow for parameter uncertainty can sometimes be incorporated into the PMSE but the correction is usually of order $1/N$ and is often omitted.

This approach assumes, not only that the correct model has been fitted, but also that 'errors' are normally distributed. It is important to check that the normality assumption is at least approximately true. The normality assumption for forecast errors also implicitly assumes that unconditional forecasting results are used. The arguably more correct use of conditional results is rarely utilized.

(ii) The distinction between a forecasting *method* and a forecasting *model* should be borne in mind. The former may, or may not, depend (explicitly or implicitly) on the latter. For large groups of series, an ad hoc forecasting method is sometimes selected. Then P.I. formulae are sometimes based on the model for which the method is optimal, but this should only be done after appropriate checks on the one-step-ahead forecasting errors. Conversely it seems silly to derive P.I.s from a model for which a given method is known *not* to be optimal, as sometimes happens.

Formulae are also now available for the Holt-Winters method which are based, not on a model, but on assuming that the method is optimal in the sense of giving uncorrelated one-step-ahead errors. When seasonality is multiplicative, these results have the interesting property that the PMSE does not necessarily increase with the lead time, such behaviour being typical of non-linear systems.

(iii) Various 'approximate' formulae for calculating P.I.s can be very inaccurate and should not be used.

(iv) When theoretical formulae are not available, or there are doubts about model assumptions, the use of empirically based or resampling methods should be considered as a general-purpose alternative. More research on this sort of approach is to be encouraged. In particular, methods for smoothing the values of the h-steps-ahead forecast standard deviations need to be found for empirically based methods, while clearer guidance is needed on how to resample time-series data.

(v) P.I.s tend to be too narrow in practice for a variety of reasons, not all of which can be foreseen. The most important reasons why out-of-sample forecasting ability is typically worse than within-sample fit is that the wrong model may have been identified or may change through time. More research is needed on whether it is possible to allow for model uncertainty. Rather than systematically widen all intervals as in (7.7.1), my current recommendation is that (7.5.1) should generally be used but that a 90% (or 80%) interval should be calculated and that the implied assumptions should be made more explicit. In particular, (7.5.1) assumes that a model has been identified correctly, that the innovations are normally distributed and that the future will be like the past.

(vi) The example in Section 7.8 demonstrates the difficult but overriding importance of model specification. In particular, for a series which is 'nearly non-stationary', it may be hard to distinguish between models which give stationary or non-stationary outcomes. Yet the difference between the limiting PMSE which results (finite for a stationary process and unbounded for a non-stationary process) is critical.

(vii) Rather than compute P.I.s based on a single 'best' model (which may be wrongly identified or may change through time), it may be worth considering a completely different approach based either on using a mixture of models, by means of Bayesian model averaging, or on using a forecasting method that is not model-based but is deliberately designed to be adaptive and robust.

(viii) Perhaps the main message of this chapter is that it is generally wise to compute P.I.s, rather than rely solely on point forecasts. However, the analyst should not trust the results blindly, and, in particular, should spell out the assumptions on which any P.I. is based.

CHAPTER 8

Model Uncertainty and Forecast Accuracy

Much of the time-series literature implicitly assumes that there is a *true model* for a given time series and that this model is *known before it is fitted to the data and then used to compute forecasts*. Even assuming that there is a true model (a big IF!), it is rarely, if ever, the case that such a model will be known a priori and there is no guarantee that it will be selected as the best fit to the data. Thus there is usually considerable uncertainty as to which model should be used, as noted earlier in Sections 3.5 and 7.7. This book has adopted the standpoint that a 'true' model usually does not exist, but rather that the analyst should try to find a model that provides an adequate approximation to the given data for the task at hand. The question then is how uncertainty about the model will affect the computation of forecasts and estimates of their accuracy, and that is the theme of this chapter.

8.1 Introduction to model uncertainty

There are typically three main sources of uncertainty in any statistical problem:

(a) Uncertainty about the structure of the model;

(b) Uncertainty about estimates of the model parameters, assuming the model structure is known;

(c) Uncertainty about the data even when the model structure and the values of the model parameters are known. This will include unexplained random variation in the observed variables, as well as measurement and recording errors.

In this chapter, the focus is on component (a), which may succinctly be described as *model uncertainty*. It is possible to further partition this component into some of the many distinct reasons why model uncertainty may arise. For example, two distinct reasons are (i) the structure of the model is misspecified a priori, and (ii) the model parameters are assumed to be fixed when, in fact, they change through time. In time-series forecasting, there is particular interest in a third source of uncertainty, namely (iii) that arising from formulating, fitting and testing a model using the *same* set of data. Clements and Hendry (1998a) provide a detailed alternative taxonomy of uncertainty as applied to forecast errors.

However uncertainty is partitioned, it should become clear to the reader that doubts about the model may sometimes have a more serious effect on forecast accuracy than uncertainty arising from other sources, as the following example will demonstrate.

Example 8.1. Fitting an AR model. Suppose we think that an AR model is appropriate for a given time series. It is quite common to simply assume that a suitable model for the mean-corrected data is first order, namely

$$X_t = \phi X_{t-1} + Z_t \qquad\qquad (8.1.1)$$

using the notation of Section 3.1.1. If we were to fit an AR(1) model, theory tells us about the errors resulting from having an estimate of the autoregression coefficient, $\hat{\phi}$, rather than its true value, ϕ. Theory also tells us about the effect of the innovations $\{Z_t\}$ on prediction intervals. However, it is much more difficult for theory to tell us about the effects of misspecifying the model in the first place. For example, suppose that the true model is AR(2), but that we mistakenly fit an AR(1). What effect will this have? Alternatively, what happens if we fit completely the wrong type of model, as, for example, if non-linear terms should be included, or the error variance is not constant through time, or we should have included appropriate trend and seasonal terms? Then the forecasts from an AR(1) model would clearly be seriously biased and estimates of forecast accuracy based on the AR(1) model would be inappropriate.

A rather different possibility is that the autoregressive term at lag one has not been assumed a priori, but rather has been found by some sort of subset selection approach. For example, the analyst might initially allow the model to include AR terms from lags one to five, and then carry out a series of tests to see which lags to include. If the term at lag one is the only one found to be statistically significant, what effect will choosing the model in this way have on subsequent inference and prediction? It is clear that a series of tests need not lead to the correct model being selected, but even if it does, the resulting estimates of model parameters may be biased by the model selection procedure, and we will see that this is indeed the case.   □

Despite the obvious importance of model uncertainty, there is not much help in the statistical literature on the implications in regard to forecasting, although some research findings are summarized in Section 8.4. In particular, they help to explain why prediction intervals tend to be too narrow, as discussed earlier in Section 7.7. Various ways of coping with model uncertainty will be reviewed in Section 8.5, but we first discuss, in Section 8.2, some general issues in time-series model building, especially the issues raised by data-driven inference. Then Section 8.3 presents two extended examples.

8.2 Model building and data dredging

A key ingredient of time-series analysis, and hence of forecasting, is the selection of a plausible approximating model that represents a given set of time series data with adequate precision for the task at hand. Some general remarks on the whole process of time-series model building were made in Section 3.5. We saw that *fitting* a particular known time-series model to a given set of data is usually straightforward nowadays, thanks to a wide range of computer software. Packages also typically carry out a range of routine *diagnostic checks* such as calculating the autocorrelation function of the residuals. In contrast, *formulating* a sensible time-series model for a given set of data can still be difficult, and this will be the prime focus of this section.

It is the exception, rather than the rule, for a model to be specified exactly from external subject-matter considerations, though a general class of models may have been suggested. As a result, it is common practice for most time-series models to be determined empirically by looking at the given time series data in an iterative, interactive way. This is exemplified by the Box-Jenkins model-building procedure, which involves an iterative cycle of model specification, model fitting, and model checking. Originally proposed by Box et al. (1994, Section 1.3.2) for building an ARIMA model, a similar iterative approach has since become standard for selecting an appropriate candidate from other classes of time-series model, and also more generally for model building in other areas of statistics. Example 8.2 outlines a typical application.

Example 8.2. A typical forecasting application. Suppose that a statistician is given monthly sales data for the past five years for a particular product and asked to make forecasts for up to 12 months ahead in order to plan production. How would the statistician go about this task?

There is no simple answer in that all decisions depend on the *context*, but the analyst will typically entertain a plausible family of possible models, such as the ARIMA class, look at a time plot of the data and at various diagnostic tools, such as the autocorrelation function, and then try appropriate models within the chosen family. A 'best' model will be selected in some way, and then inferences and forecasts will be computed conditional on the selected model being 'true'. This is done even though the model has actually been selected from the same data which are now being (re-)used to make predictions. The question is "Is this approach sound?"

Suppose, for example, that the time-series analyst starts the analysis by entertaining the class of ARIMA(p, d, q) models for say $0 \leq p, d, q \leq 2$. This looks fairly innocent, but actually allows a total of 27 possible models. Moreover, when the analyst looks at the data, there are additional possibilities to consider such as finding seasonality (which suggests trying a seasonal ARIMA model even though such a model was not entertained initially) or finding non-linearities (which would suggest a completely different class of models). This all means that the analyst effectively looks at many different models. Nevertheless the analyst typically ignores the effect of having tried many models when it comes to making forecasts. □

Most time-series forecasting is carried out conditional on the best-fitting model being true. But should it be? Many years ago it may have been normal to fit a *single* model to a given set of data. Nowadays the increase in computing power has completely changed the way in which time-series analysis is typically carried out (not necessarily for the better!). A model is often selected from a wide class of models by optimizing a statistic such as the AIC. As well as choosing from a wide class of models, the data-analysis procedure may also involve strategies such as (i) excluding, down-weighting or otherwise adjusting outliers; (ii) transforming one or more variables to achieve normality and/or constant residual variance. As a result of all this, the analyst may in effect consider tens, or even hundreds of models, either explicitly or sometimes implicitly (it is not always clear just how many models are considered).

Having selected a model, the analyst typically proceeds to estimate the parameters of this best-fitting model using the *same* techniques that would be used in traditional statistical inference where the model is assumed known a priori. The properties of forecasts from the resulting model are also generally calculated as if the model were known in the first place. Thus the standard analysis does not take account of the fact that (i) the model has been selected from the same data used to make inferences and predictions and (ii) the model may not be correct anyway. Thus the standard analysis ignores model uncertainty. Unfortunately, this is "logically unsound and practically misleading" (Zhang, 1992). In particular, least squares theory is known not to apply when the same data are used to formulate *and* fit a model, as happens routinely in time-series analysis. Example 8.3 below illustrates this point for a particular situation.

Statisticians have typically ignored this type of problem, partly because it is not clear what else could or should be done. Little theory is available for guidance, and the biases which result when a model is formulated and fitted to the *same* data are not well understood. Such biases can be called *model-selection biases* (Chatfield, 1995b).

8.2.1 Data dredging

When a model is selected from a large set of candidate models, and then used for inference and prediction, statisticians often refer to the general process as *data dredging* or *data mining* – see Section 3.5.2. *Data snooping* is another possible description. In fact, the term data mining is now used by computer scientists in a rather different way to mean *knowledge discovery in very large databases*. Thus it will be prudent to avoid the phrase 'data mining', and so we opt for 'data dredging'. An alternative description of data dredging is as a form of *data-driven inference*. Although widely practiced, data dredging is rightly seen as a rather suspect activity, partly because over-fitting may occur, and partly because the analyst may not know how to allow for the effect of formulating the model from the data prior to using the fitted model. Thus the term is often used in a rather derogatory way.

In econometrics, there is much activity that might be classified as data dredging, although some users may not realize what they are doing. One example is the search for calendar effects in stock market behaviour. If you look hard enough at lots of financial series over many different time periods, the analyst will eventually be able to spot some apparently interesting regularity, even if the effect is actually spurious and fails to generalize to other time periods or to other similar measured variables. For example, all sorts of rules have been suggested for investing in the stock market, such as 'Sell on the first day of the month' or 'Buy on Friday'. Tested in isolation, such effects may well appear to be highly significant. However, the effect has only been tested because it has been spotted in the data. In other words, the hypothesis has been generated by the same data used to test it. Sadly, this sort of data-driven inference is likely to produce spurious results. If the analyst takes into account all the other rules which have implicitly been considered and rejected, then the significance of the interesting-looking effects is highly diminished. There are literally thousands of rules which might be considered and Timmermann (1998) uses bootstrapping to show that many calendar effects are not significant when the hypothesis-generation process is taken into account. This is, of course, what would be expected for a properly performing market.

In econometrics, the difficulties in model building are further complicated by the common practice of pretesting various hypotheses, such as testing for a unit root, for autocorrelated residuals, or for the presence of a change point. The desire to carry out such tests indicates awareness of model uncertainty, but is viewed with suspicion by many statisticians, especially when a large number of tests is performed – see the remarks in Section 3.5.2. Why, for example, should the presence of a unit root be taken as the null hypothesis when examining the possible presence of trend (see Section 3.1.9 and Chatfield, 1995b, Example 2.5)? We do not pursue these matters here except to note that (i) inference following model testing is biased, (ii) testing implicitly assumes the existence of a true model that is included in the set of models entertained,[1] (iii) the more models that are entertained, and the more tests that are carried out, the lower the chance of choosing the 'correct' model. In fact, most model-builders would admit (privately at least!) that they do not really believe there is a true model (see Chatfield, 1995b, Section 3.1). Rather, a model is seen as a useful description of the given data which provides an adequate approximation for the task at hand. Here the *context* and the *objectives* are key factors in making such a judgement.

---

[1] Note that this is called the *M-closed view* by Bernado and Smith (1994, Section 6.1.2).

8.3 Examples

Given the difficulty of finding general theoretical results about the effects
of model selection on subsequent inference, the use of specific examples,
perhaps employing simulation, can be enlightening. This section presents
expanded versions of the two main examples in Chatfield (1996b) to show
that data-dependent model-specification searches can lead to non-trivial
biases, both in estimates of model parameters and in the forecasts that
result.

Further examples are given by Miller (1990) and Chatfield (1995b,
Examples 2.2 and 2.3) on regression modelling, and by Hjorth (1987,
Examples 5 and 7), Hjorth (1994, Example 2.2) and Chatfield (1995b,
Examples 2.4 and 2.5) on time-series analysis and forecasting.

Example 8.3. Fitting an AR(1) model. Consider the first-order
autoregressive (AR(1)) time-series model, namely:

$$X_t = \phi X_{t-1} + Z_t \tag{8.3.1}$$

where $|\phi| < 1$ for stationarity and $\{Z_t\}$ are uncorrelated $N(0, \sigma_Z^2)$ random
variables. Given a sample of mean-corrected data, it is straightforward to
fit the above AR(1) model and estimate ϕ. However, in practice with real
data, the analyst is unlikely to know a priori that the AR(1) model really
is appropriate, and will typically look at the data first to see what model
should be adopted.

A simple (perhaps oversimplified) identification procedure consists of
calculating the first-order autocorrelation coefficient, r_1, and fitting the
AR(1) model if, and only if, r_1 is significantly different from zero. What
this procedure does is to eliminate the possibility of getting small values
of $\hat{\phi}$ which correspond to small values of r_1. Instead, large values of $\hat{\phi}$ are
more likely to occur, as can readily be demonstrated theoretically or by
simulation. Then the properties of the resulting estimate of ϕ should be
found, not by calculating the unconditional expectation of $\hat{\phi}$, but rather
the conditional expectation $E(\hat{\phi}|\ r_1$ is significant). It is intuitively clear
that the latter will not equal ϕ so that a bias results. This bias can be
substantial as demonstrated by Chatfield (1995b, Example 2.4) for the
case where time series of length 30 were simulated from an AR(1) model
with $\phi = 0.4$. Then it was found that $E(\hat{\phi}|\ r_1$ is significant) is nearly 0.5,
which is substantially larger than the true value of 0.4.

This example emphasizes that, in assessing bias, the analyst must be
clear exactly what the inference is conditioned on. Theory tells us about
the properties of the unconditional expectation of $\hat{\phi}$ where it is assumed
that an AR(1) model is always fitted to a given set of data. However, when
the model is selected from the data, the model-selection process needs to
be taken into account.

There is also a third estimator which may arguably be preferred to the unconditional estimator, $\hat{\phi}$, namely

$$\hat{\phi}_{PT} = \left\{ \begin{array}{ll} \hat{\phi} & \text{if } r_1 \text{ is significant} \\ 0 & \text{otherwise} \end{array} \right.$$

This estimator can be recognized as a simple example of what econometricians call a *pre-test* estimator (e.g. Judge and Bock, 1978). It arises by recognizing that, when r_1 is not significant and an AR(1) model is not fitted, this could be regarded as fitting an AR(1) model with $\phi = 0$. It is immediately apparent that the three quantities $E(\hat{\phi})$, $E(\hat{\phi} \,|\, r_1$ is significant) and $E(\hat{\phi}_{PT})$ will generally not be equal, and will, moreover, have different sampling distributions and hence different variances. In any case, the expectations will depend on the actual estimation procedure used. For example, in the case considered above, where series length 30 were simulated from an AR(1) with $\phi = 0.4$, it was found that the Yule-Walker unconditional estimator (based on r_1) gave $\hat{\phi} = 0.32$, which is seriously biased, whereas the MINITAB package, using a non-linear routine, gave $\hat{\phi} = 0.39$, which is nearly unbiased. When series with non-significant values of r_1 are excluded, it was found that $E(\hat{\phi} \,|\, r_1$ is significant) is about 0.5, while if excluded series are counted as AR(1) series with $\phi = 0$, then $\hat{\phi}_{PT}$ is about 0.3. These two conditional expectations are about equidistant from the true value of 0.4.

Given that estimators of model coefficients are biased, it is not surprising to find that estimates of the residual standard deviation are also likely to be biased (see further remarks on this point in Section 8.4).

By now, the reader may be wondering which estimator to use, and it is hard to offer general advice on this. The choice of 'best' estimator may well depend on the context. We usually prefer estimators to be unbiased and to have the smallest possible mean square error (MSE), but correcting bias may actually increase MSE. What we can say is that, if a model is selected using a given set of data, then this should be recognized when computing estimators from the same data.

Of course, the above model-selection procedure is simpler than would normally be the case in time-series analysis. More typically the analyst will inspect autocorrelations and partial autocorrelations of a suitably differenced series, allow the removal or adjustment of outliers and entertain all ARIMA models up to say third order. Choosing a 'best' model from such a wide set of possibilities seems likely to make model selection biases even larger.

While this example has focussed on estimates of model parameters, the results are of course relevant to forecasting since point forecasts and prediction intervals are calculated conditional on the fitted model. If estimates of model parameters are biased, then the resulting forecasts can also be expected to be biased, and this is indeed the case, as will be illustrated below.

Suppose, for example, that a time series of length 30 is taken from a zero-mean AR(1) process with $\phi = 0.4$ and $\sigma_Z = 1$. Further suppose we only fit the AR(1) model if r_1 is significantly different from zero. Then, as noted earlier, the conditional expectation of $\hat{\phi}$ is about 0.5, rather than 0.4. Then the one-step-ahead point forecast, $\hat{x}_{30}(1) = \hat{\phi} x_{30}$, will have a conditional expectation equal to $0.5 \times x_{30}$ rather than $0.4 \times x_{30}$. This is a 25% bias. What effect will this have on the accuracy of the forecast, as measured by prediction mean square error (PMSE)? If the true model is known, the variance of the correct one-step-ahead forecast is unity, so that the PMSE (= variance + square bias) of the biased forecast is $(1 + 0.1^2 x_{30}^2)$. Thus, although the bias is quite large, the effect on the accuracy of point forecasts is minimal except when x_{30} is large, say bigger than about 2. However, it should be remembered that this AR(1) model, with a relatively low value for ϕ, has poor forecasting ability anyway, since it explains less than 20% of the total variance $(\sigma_X^2 = \sigma_Z^2 / (1 - \phi^2) = 1.19 \sigma_Z^2$. For models with better intrinsic forecasting accuracy, the effect of biased parameters is likely to have a greater proportionate effect on forecast accuracy.

As regards prediction intervals, the estimate of residual variance is also likely to be biased but in a downward direction – see Section 8.4. A bias of 10% in $\hat{\sigma}_Z$ is probably conservative. Taking $x_{30} = 1.5$ for illustrative purposes, the true 95% prediction interval should be $0.4 \times 1.5 \pm 1.96 \times 1$, while the estimated interval will on average be $0.5 \times 1.5 \pm 1.96 \times 0.9$. Thus the true interval is $(-1.36, 2.56)$ but the estimated interval will on average be $(-1.01, 2.51)$. These intervals are non-trivially different.

Unfortunately, there appears to be no easy general way to quantify the theoretical extent of these biases and progress seems likely to be made primarily by simulation and by empirical experience.   □

Example 8.4. Fitting neural network models. Neural network (NN) models were introduced in Section 3.4.4. They can be thought of as a type of non-linear regression model, and have recently been applied to various time-series forecasting problems. NNs comprise a large class of models which allows the analyst to try many different architectures with different numbers of hidden layers and units. The parameters of an NN model are the weights which indicate the strength of particular links between the input variables and the different hidden units or between the hidden units and the output(s). Thus NN models typically have a (much) larger number of parameters than time-series models. When this feature is combined with the large number of models which can be tried, a good (within-sample) fit can usually be obtained with an NN model. However, there is a real danger of *overfitting*, and the forecasting ability of NNs is still uncertain – see Section 6.4.2.

Faraway and Chatfield (1998) presented a case study where they applied a variety of NN models to forecasting the famous airline data which was plotted in Figure 2.1. The accuracy of the out-of-sample forecasts was generally disappointing. The relationship between in-sample and out-of-sample forecast accuracy was further investigated by Chatfield (1996b,

Example 2). This example summarizes the earlier results and also extends the work in two ways; first, by giving more attention to the bias-corrected version of AIC, denoted by AIC_C, which was introduced in Section 3.5, and second, by demonstrating the clarification introduced by standardizing tabulated values of information criteria statistics.

All NN models fitted by Faraway and Chatfield (1998) were of the usual feedforward type with one hidden layer of neurons. The input variables were the values of the given variable (the number of airline passengers) at selected lags so that attention was restricted to univariate forecasts where forecasts of X_t depend only on past values of the series. The output variables were the forecasts for different horizons. The logistic activation function was used at each neuron in the hidden layer and the identity activation function was used at the output stage. Initially the models were fitted to the first eleven years of data (the training set in NN jargon) and the data in the twelfth year (the test set) was used for making genuine out-of-sample forecast comparisons. There were many practical problems in practice, such as avoiding local minima and choosing sensible starting values, but the reader is referred to Faraway and Chatfield (1998) for details.

We divide our comments on the results into two parts, namely (i) those concerned with selecting an NN model, and (ii) those concerned with evaluating the out-of-sample forecasts.

Model selection. When fitting NN models, the data summary tools, such as correlograms, have less value than with other classes of time-series model because the theoretical functions for different NN models are not available. However, the time plot and correlogram might help to give some indication as to what lagged variables should be included as inputs (e.g. include lag 12 for seasonal data having large autocorrelation at lag 12). There seems little scope for using tests to help identify an NN model, and so this analysis was based primarily on looking at the model-selection criteria introduced in Section 3.5. By inspecting the time plot and sample correlograms, it seems clear that, in order to avoid silly NN models, the values at lags one and twelve should always be included as input variables. Some other NN models (e.g. using values at lags one to four only as input variables) were tried just to make sure that they did indeed give poor results. Some selected results, taken from Chatfield (1996b), are shown in Table 8.1. The quantities tabulated (and the notation adopted) are as follows:

1. The 'lags' shows the lag values of the input variables used. The number of input variables is denoted by k.

2. The number of neurons selected for the hidden layer. This is denoted by H. We use the notation $\text{NN}(j_1, j_2, \ldots j_k; H)$ to denote the NN model having input variables at lags $j_1, j_2, \ldots j_k$, and with H neurons in the hidden layer.

lags	no. of hidden neurons (H)	no. of params. (p)	$\hat{\sigma}$	Measures of Fit	
				AIC	BIC
1,12,13	1	6	0.102	−537.1	−514.4
1,12,13	2	11	0.098	−543.1	−501.4
1,12,13	4	21	0.093	−546.8	−467.4
1,12	2	9	0.144	−456.3	−422.2
1,12	4	17	0.145	−447.7	−383.5
1,12	10	41	0.150	−423.7	−268.4
1,2,12	2	11	0.141	−459.4	−417.7
1,2,12	4	21	0.139	−454.6	−375.1
1,2,12,13	2	13	0.097	−543.5	−494.4
1,2,12,13	4	25	0.093	−543.1	−448.7
1−13	2	31	0.091	−544.8	−427.6
1−13	4	61	0.067	−605.1	−374.6

Table 8.1 *Some measures of fit for various NN models for the airline data using the first 132 observations as the training set*

3. The 'no. of pars.' gives the number of weights (parameters) which have to be estimated. This is given by $p = (k + 2)H + 1$.

4. $\hat{\sigma}$ = estimated residual standard deviation for the training set. This was calculated as $\sqrt{[S/(N - p)]}$ where S denotes the sum of squared residuals over the training period and N denotes the number of observations in the training set.

5. AIC = Akaike's information criterion. This was approximated by $N \ln(S/N) + 2p$.

6. BIC = the Bayesian information criterion. This was approximated by $N \ln(S/N) + p + p \ln N$.

As k and H are increased, the number of parameters (p) increases alarmingly. Several models have in excess of 20 parameters even though the number of observations in the training set is only 132. Many (most?) analysts would guess that it would be unwise to have more than about 10 parameters with so few observations. Generally speaking, Table 8.1 demonstrates that the more parameters are fitted, the lower will be the value of $\hat{\sigma}$, as would be expected. Thus these values tell us very little and could even lead the analyst to wrongly choose a model with too many parameters. Instead, we look at some model-selection criteria which penalize the addition of extra parameters. The values of AIC are tabulated but are hard to read from Table 8.1 as presented. It takes some effort to search for, and find, the minimum value of AIC which occurs for the 61-parameter NN(1−13; 4) model. However, it is very doubtful whether any experienced statistician would seriously entertain a model with so many parameters. The model which gives the next lowest AIC, namely NN$(1, 12, 13; 4)$, still has 21 parameters. In contrast, the model giving the lowest BIC (which penalizes extra parameters more severely than the AIC) has just six parameters, namely NN$(1, 12, 13; 1)$. Thus the use of BIC rather than AIC leads to a completely different choice of model!

We will now give a revised version of Table 8.1 in two respects. First, the values of AIC and BIC contain an arbitrary constant. It is the *differences* between them which are of interest. The statistics are scale-independent and a useful guideline is that any model whose AIC is within about 4 of the minimum AIC is still a good model. This suggests standardizing the values so that the minimum value is at zero. This will make the values much easier to read and compare, and is recommended by Burnham and Anderson (1998). The second extension is to include values of the bias-corrected version of AIC, which was denoted by AIC_C in Section 3.5. Many statisticians are, as yet, unaware of this statistic, and most software does not tabulate it. Fortunately, it is easy to calculate (approximately) by adding $2(p+1)(p+2)/(N-p-2)$ to the AIC. This correction term is small when p is small compared with N, but can become large if p/N exceeds about 0.05. The resulting values of the (standardized) values of AIC, AIC_C and BIC are shown in Table 8.2. The notation ∇AIC, for example, denotes the difference between the AIC for a given model and the minimum value of AIC over all models. Of course, some care is needed if new models are added during the analysis resulting in a lower minimum value. All standardized values would need to be recalculated by differencing with respect to the new minimum.

This table is much easier to read. As well as making it easier to pick the 'best' model corresponding to each statistic, the table also helps to get a *ranking* of models – which models are 'best' or 'nearly best', which are clearly inferior and which are intermediate. The AIC_C is midway between the AIC and BIC in its effect on the size of the model chosen as it chooses the 11-parameter NN$(1, 12, 13; 2)$ model. There are three other models within 4 of the minimum value of AIC_C, including that chosen by the BIC. However, the model chosen by the (ordinary) AIC has AIC_C about 49 above the minimum value and this suggests it is a poor model.

There is no single obvious choice as the 'best' model based on all these measures of fit. The models selected by AIC_C and by BIC both seem plausible, as do one or two other models. What is clear is that the model selected by the ordinary AIC is not acceptable. On the basis of Table 8.2, two recommendations can be made:

1. Values of AIC-type statistics should be standardized to have a minimum value of zero in order to make them easier to compare.

2. The bias-corrected version of AIC, namely AIC_C, is preferred to the ordinary AIC.

One general remark is that model uncertainty is clearly high. Of course,

lags	no. of hidden neurons (H)	no. of pars. (p)	$\hat{\sigma}$	Measures of Fit ∇AIC	∇AIC$_C$	∇BIC	Notes
1,12,13	1	6	0.102	68	4	0	Best BIC
1,12,13	2	11	0.098	62	0	13	Best AIC$_C$
1,12,13	4	21	0.093	58	3	47	
1,12	2	9	0.144	149	86	92	
1,12	4	17	0.145	157	99	131	
1,12	10	41	0.150	182	158	246	
1,2,12	2	11	0.141	146	84	97	
1,2,12	4	21	0.139	150	95	139	
1,2,12,13	2	13	0.097	62	1	20	
1,2,12,13	4	25	0.093	62	11	66	
1−13	2	31	0.091	60	17	87	
1−13	4	61	0.067	0	49	140	Best AIC

Table 8.2 *Some standardized measures of fit for various NN models for the airline data using the first 132 observations as the training set*

some models may be discarded as having high values for all the model-selection criteria. For example, one silly model we fitted, namely the NN$(1, 2, 3, 4; 2)$ model, gave very high values for AIC, AIC$_C$ and BIC as did the NN$(1, 12; 10)$ model in Table 8.2. However, several other models are close competitors to those selected as 'best' by AIC$_C$ or BIC. If forecasts, and especially forecast error variances, are calculated conditional on just one selected model, the results will not reflect this uncertainty.

One final remark is that we have confined attention to NN models in the discussion above. However, it could be worth trying an alternative family of models. In fact, results for a good Box-Jenkins ARIMA model (the so-called 'airline' model) gave measures of fit which were better than the best NN model in terms of both AIC$_C$ and BIC.

Forecast accuracy. We now compare forecast accuracy for the different NN models. As always, we look at out-of-sample forecasts as the critical issue in applying NN models (or any data-driven model) is the performance on data *not* used for training the model. Table 8.3 presents three measures of forecast accuracy. SS_{MS} and SS_{1S} are the sums of squares of multi-step and one-step-ahead (out-of-sample) forecast errors for the last year's data. The multi-step forecasts were all made in month 132. The one-step-ahead forecasts were made by forecasting one step ahead from month 132, then bringing in observation 133 so as to forecast observation 134, and so on to the end of the series. In all, there are twelve one-step-ahead forecasts for

lags	no. of hidden neurons (H)	no. of params. (p)	Fit $\hat{\sigma}$	Measures of Forecast Accuracy			Notes
				SS_{MS}	SS_{1S}	$\hat{\sigma}_{pred}$	
1,12,13	1	6	0.102	0.33	0.51	0.20	Best BIC
1,12,13	2	11	0.098	0.33	0.50	0.20	Best AIC$_C$
1,12,13	4	21	0.093	0.54	0.62	0.23	
1,12	2	9	0.144	0.35	0.34	0.17	
1,12	4	17	0.145	0.38	0.44	0.19	
1,12	10	41	0.150	0.51	0.59	0.22	
1,2,12	2	11	0.141	0.34	0.29	0.16	
1,2,12	4	21	0.139	6.82	1.03	0.29	
1,2,12,13	2	13	0.097	0.37	0.52	0.21	
1,2,12,13	4	25	0.093	0.34	0.52	0.21	
1 − 13	2	31	0.091	1.08	0.71	0.24	
1 − 13	4	61	0.067	4.12	1.12	0.31	Best AIC

Table 8.3 *Some measures of forecast accuracy for various NN models for the airline data using the last 12 observations as the test set*

each model. The prediction error standard deviation, denoted by $\hat{\sigma}_{pred}$, is estimated by $\sqrt{(SS_{1S}/12)}$. In order to compare (out-of-sample) forecast accuracy with (within-sample) fit, Table 8.3 also gives the fit residual standard deviation, $\hat{\sigma}$, for comparison with σ_{pred}.

Table 8.3 tells us that getting a good fit – meaning a low value of $\hat{\sigma}$ – is a poor guide to whether good predictions result. Indeed models with a *smaller* numbers of parameters generally give *better* (out-of-sample) predictions even though they may appear to give a worse fit than less parsimonious models. In particular, the model selected as 'best' by AIC$_C$ or by BIC gives much better out-of-sample predictions than the model selected as 'best' by AIC. This finding confirms the recommendation that AIC$_C$ and BIC are better model-selection criteria than AIC.

Table 8.3 also allows us to compare fit with forecast accuracy more generally. The results might be surprising to the reader with little experience of making genuine out-of-sample forecasts. The within-sample estimate of the error standard deviation (i.e. $\hat{\sigma}$) is typically much less than the (out-of-sample) one-step-ahead prediction error standard deviation (i.e. $\hat{\sigma}_{pred}$), whereas if we have selected the true model, we would expect these two statistics to be estimates of the same quantity. For the better models (with a small number of parameters and low AIC$_C$ or BIC), we find $\hat{\sigma} \simeq 0.1$ and $\hat{\sigma}_{pred} \simeq 0.2$. So the latter is about double the former. For the more dubious models (with low $\hat{\sigma}$ and low AIC, but with higher numbers of parameters), the ratio of $\hat{\sigma}_{pred}$ to $\hat{\sigma}$ becomes disturbingly large – for example, it rises to 4.6 for the 61-parameter model with the best AIC.

The above results report what happens when different NN models are fitted to the airline data using the first 132 observations as the training set. Qualitatively similar results were found using the first 126 observations as the training set and also using a completely different data-set, namely the Chatfield-Prothero sales data (Chatfield and Faraway, 1996). It was also found that the relatively poor accuracy of out-of-sample forecasts (as compared with within-sample fit) is not confined to NN models. For example, when Box-Jenkins seasonal ARIMA models were fitted to the airline data, it was also found that $\hat{\sigma}_{pred}$ was typically at best twice as large as $\hat{\sigma}$ as for NN models.   □

Why is $\hat{\sigma}_{pred}$ so much larger than $\hat{\sigma}$ in Example 8.4, and is this a finding which generalizes to other data sets and models? The answer to the second question seems to be "Yes" judging by the empirical evidence which is reviewed by Fildes and Makridakis (1995) and Chatfield (1996b) – see Section 7.7. *Out-of-sample forecast accuracy is generally (much) worse than would be expected from within-sample fit.* Some theoretical results, such as the optimism principle (see Section 8.4), help to explain the above. While there are other contributory factors, it seems likely that model uncertainty is the prime cause. Either an incorrect model is identified or the underlying model is changing through time in a way which is not captured by the forecasting mechanism. This gives further justification to our earlier recommendation (Chapter 6) that comparisons of different forecasting models and methods should be made on the basis of out-of-sample predictions.

8.4 Inference after model selection: Some findings

This section reviews previous research on model uncertainty and, in particular, on data dredging, which here means inference on a given set of data during which the same data are used both to select an appropriate model and then to test, fit and use it. Chatfield (1995b) has given a general review of work in this area up to 1995. Other general references include Hjorth (1994), Burnham and Anderson (1998, Chapter 4) and, in the econometrics literature, Clements and Hendry (1998a, especially Chapters 7 and 11).

Assuming that a 'true' model does not usually exist, the analyst looks for a model which provides an adequate approximation for the task at hand. However, if we search widely for a best-fitting model, but then carry out inferences and prediction as if we believe that the best-fitting model is true, then problems inevitably arise. It seems illogical to effectively admit that we do not know what model is appropriate by searching for a 'best' model, but then ignore this uncertainty about the model in subsequent inferences and predictions. Unfortunately, it is not easy to specify how we should proceed instead.

Statistical theory has generally not kept pace with the computer-led revolution in statistical modelling practice and there has been rather little progress in understanding how inference is affected by prior model-selection. It is now known that large biases can arise when model parameters are estimated from the same data previously used to select the model, but there is no simple way of deciding how large these biases will be. One key message is that *the properties of an estimator may depend, not only on the selected model, but also on the selection process.* For this reason, the analyst who reports a best-fitting model should also say how the selection was made and, ideally, report the full set of models that have been investigated. In practice, it can be difficult to comply with the latter requirement, as many models may be investigated implicitly, rather than explicitly.

The use of a model-selection statistic essentially partitions the sample space into disjoint subsets, each of which leads to a different model. This vantage point enables the derivation of various inequalities regarding the expectation of the optimized statistic and provides a theoretical justification for what Picard and Cook (1984) call the *Optimism Principle*, namely that the fitting of a model typically gives optimistic results in that performance on new data is on average worse than on the original data. In particular, if a time-series model is selected by minimizing the within-sample prediction mean square error (PMSE), then the Optimism Principle explains why the fit of the best-fitting model is typically better than the resulting accuracy of out-of-sample forecasts. This is reminiscent of the *shrinkage* effect in regression (e.g. Copas 1983), and of experience with discriminant analysis where discrimination on a new set of data is typically worse than for the data used to construct the discrimination rule.

8.4.1 Prediction intervals are too narrow

One important consequence of formulating and fitting a model to the same data is that *prediction intervals are generally too narrow.* This phenomenon is well documented, and empirical studies (see Section 7.7) have shown that nominal 95% prediction intervals will typically contain (much) less than 95% of actual future observations. A variety of contributory reasons have been suggested for this effect, but model uncertainty is arguably the most important. If the best-fitting model is chosen from many alternatives, the residual variance, and hence the prediction mean square error (PMSE), are typically underestimated. Moreover, the wrong model may be identified, or the model may change through time.

Although prediction intervals tend to be too narrow, there is an alarming tendency for analysts to think that narrow intervals are somehow better than wider ones, even though the latter may well reflect model uncertainty more realistically. This is illustrated by Draper's (1995) example on forecasting the price of oil, as discussed earlier in Section 7.5.7. Draper entertained a variety of models and scenarios, giving a wide range of point forecasts. Sadly, the point forecasts and prediction intervals were all well away from the actual values that resulted. A model uncertainty audit suggested that only about 20% of the overall predictive variance could be attributed to uncertainty about the future of oil prices conditional on the selected model and on the assumptions (the scenario) made about the future. Thus prediction intervals, calculated in the usual way, will be too narrow. Yet the variance conditional on the model and on the scenario is all that would normally be taken into consideration when calculating prediction intervals. Other case studies (e.g. Wallis, 1986), which have examined the decomposition of forecast errors, have also found that the contribution of uncertainty in model specification to predictive variance can be substantial.

Model uncertainty is increasingly important for longer lead times where prediction intervals, conditional on a single model, are likely to be much too narrow. It is easy to demonstrate that models which are mathematically very different may be virtually indistinguishable in terms of their fit[2] to a set of data and in the short-term forecasts which result, but may give very different long-term predictions. This is illustrated in the following example.

Example 8.5. For some near-non-stationary sets of data, an AR(1) model with a parameter close to unity will give a similar fit and similar one-step-ahead forecasts to those from a (non-stationary) random walk model. However, the multistep-ahead forecasts from these two models are quite different, especially in terms of their accuracy. For simplicity, we consider an AR(1) model with zero mean, namely

$$X_t = \phi X_{t-1} + Z_t \qquad\qquad (8.4.1)$$

where $|\phi| < 1$ for stationarity and $\{Z_t\}$ are uncorrelated $N(0, \sigma^2)$ random variables. Given mean-corrected data to time N, the one-step-ahead forecast is $\hat{x}_N(1) = \phi x_N$, with forecast error variance σ^2. For comparison, we look at the random walk model given by

$$X_t = X_{t-1} + Z_t \qquad\qquad (8.4.2)$$

This model could be regarded as a non-stationary AR(1) model with $\phi = 1$. Given data to time N, the one-step-ahead forecast for the random walk is $\hat{x}_N(1) = x_N$ with the same forecast error variance as for the AR(1) model. If ϕ is close to unity, the point forecasts are also very similar. However, when we look at the multistep-ahead forecasts, the picture is very different. For h steps ahead, the point forecasts for the two models are $\phi^h x_N$ and x_N, respectively. Thus, for the stationary AR(1) model, the forecasts tend to the process mean (zero in this case), while those for the random walk are constant and equal to the most recent value. The discrepancy between

---

[2] Remember that the fit is usually measured by the within-sample one-step-ahead forecast errors.

the limiting forecast error variances is even more startling. For the AR(1) model, the forecast error variance is given by $\sigma^2(1 - \phi^{2h})/(1 - \phi^2)$. As h increases, this tends to a finite value, namely $\sigma^2/(1 - \phi^2)$ which is the variance of the AR(1) process. However, for the random walk model, the forecast error variance is $h\sigma^2$. This increases without bound as h increases. More generally, as noted in Section 7.8, the forecast error variance for a stationary model tends to a finite value as h increases, but tends to infinity for a non-stationary model.   □

Example 8.5 demonstrates that getting the wrong form of differencing may make little difference to short-term point forecasts, but can make a large difference for long-term forecasts. This means that, if there is uncertainty about the model (as there usually will be), then the effects on multi-step-ahead forecasts could be substantial. It really is a mistake to ignore model uncertainty.

    Similar remarks apply to extrapolating from any model. An instructive example concerns the Challenger space shuttle disaster data where it is hard to distinguish between several models in terms of fit, but where the long-term extrapolations are very different (Draper, 1995, Section 6.2). Forecasting the spread of AIDS provides another convincing example (Draper, 1995, reply to the discussion).

8.4.2 The results of computational studies

The difficulty in making theoretical progress has led to a number of studies, using a variety of computational procedures, including *simulation, resampling, bootstrapping*, and *cross-validation*. Only a brief summary will be given here as many studies were more concerned with parameter estimation than with forecasting per se, and early results were reviewed in some detail by Chatfield (1995b, Section 4). As one example, Miller (1990, p. 160) found alarmingly large biases, of the order of one to two standard errors, in the estimates of regression coefficients when using subset selection methods in multiple regression. Hjorth (1987, Example 5) simulated data from an ARMA(1, 1) model, but found that the correct form of ARMA model was identified in only 28 out of 500 series, using a standard identification procedure. For the other 472 series, a variety of alternative AR, MA and ARMA models were identified. The properties of the estimates of the ARMA(1, 1) model parameters for the 28 series differed greatly from those arising when the ARMA(1, 1) model was fitted to all 500 series. Furthermore (and this deserves special emphasis) the average estimated PMSE was *less than one-third* of the true PMSE for the model which was actually fitted. In other words, by allowing a choice of ARMA model, the estimate of forecast accuracy was hopelessly optimistic as compared with the true forecast accuracy of the ARMA(1, 1) model. More recently, Pötscher and Novak (1998) simulated various MA and AR models but selected the order from the data. They found that "the

distribution of post-model-selection estimators frequently differs drastically
from the distribution of LS estimates based on a model of fixed order". It
is unfortunate that results such as these continue to be largely ignored in
practice.

Some data-dependent model-selection procedures are quite complicated
but computational methods can still be used to study their effect on
subsequent inference provided the model-selection procedure is clearly
defined. For example, Faraway (1992) simulated the actions taken during
regression analysis, including the handling of outliers and transformations.
However, it is worth noting that some model-selection procedures may
involve subjective judgement and then it may not be possible to simulate
them computationally.

Apart from simulating the behaviour of model-selection procedures, it
may be possible to use other computational approaches to assess model
uncertainty. For example, Sauerbrei (1999) shows how to use resampling
and cross-validation to simplify regression models in medical statistics and
assess the effect of model uncertainty. However, as noted in Section 7.5.6,
resampling time-series data is particularly difficult because of the ordered
nature of the data and because one has to avoid conditioning on the
fitted model – otherwise any results would not reflect model uncertainty.
Nevertheless, careful bootstrapping can overcome much of the bias due to
model uncertainty.

8.4.3 Model checking

We have concentrated on inferential biases, but it should be noted that
the literature on *model checking* in time-series analysis is also suspect. In
statistical model-building, it is theoretically desirable for a hypothesis to
be validated on a second confirmatory sample. However, this is usually
impossible in time-series analysis. Rather, diagnostic checks are typically
carried out on the *same* data used to fit the model. Now diagnostic
tests assume the model is specified a priori and calculate a P-value as
Probability(more extreme result than the one obtained | model is true).
But if the model is formulated, fitted and checked using the same data,
then we should really calculate Probability(more extreme result than the
one obtained | model has been selected as 'best' by the model-selection
procedure). It is not clear in general how this can be calculated. However,
it is clear that the good fit of a best-fitting model should not be surprising,
and empirical experience suggests that diagnostic checks hardly ever reject
the best-fitting time-series model precisely because it is the best fit!

A striking example of the above phenomenon is provided by the many
attempts to find calendar effects in stock market data. As described in
Section 8.2.1, if you look hard enough at financial data, you are likely to
spot what appear to be interesting and significant regularities. However, if
tested in the light of the hypothesis-generation procedure, such effects are
usually not significant.

8.5 Coping with model uncertainty

The choice of model is crucial in forecasting as the use of an inappropriate model will lead to poor forecasts. Even after the most diligent model-selection process, the analyst should realize that any fitted model is at best a *useful approximation*. In view of the potential seriousness of model-specification error, it is usually inadequate to describe uncertainty by computing estimates of forecast accuracy which are conditional on the fitted model being true. Instead, we need to find ways of getting more realistic estimates of prediction error, perhaps based on resampling methods, or on mixing several models. More generally, we need to find ways of coping with, allowing for, or circumventing, the effects of model uncertainty on forecasts. This section discusses various alternatives to the rigidity of assuming the existence of a single known true model, and considers the implications in regard to the choice of forecasting method.

8.5.1 Choosing a single model

There are many procedures for selecting a best-fit model (see Section 3.3), but they are not our concern here. Instead, we focus on some more general questions that need to be considered when choosing a single model to describe a set of data, if that is the approach to be adopted.

Global or local model? Some forecasters aim to fit what may be described as a *global* model, where the structure of the model and the values of the model parameters are assumed to be constant through time. Regression models with constant coefficients, and growth curve models may arguably be described in this way. However, it is rather inflexible to assume that a global model is appropriate for a given time series. For example, regression models, with time as an explanatory variable, were used for many years to model trend in a deterministic way, but such models have now largely fallen out of favour, as compared with techniques which allow the trend to change through time – see Section 2.3.5. More generally, models which allow parameters to adapt through time are sometimes called *local* models. Models of this type include structural and state-space models. In particular, the dynamic linear models of West and Harrison (1997) put parameter nonconstancy to centre stage (Clements and Hendry, 1998b). Local models are often fitted by some sort of updating procedure, such as the *Kalman filter*, which is easy to apply using a computer, and the use of such techniques seems likely to increase.

In this regard it is interesting to remember that simple *exponential smoothing* (which is a very simple type of Kalman filter) is optimal for two models which appear to be of a completely different type, namely the ARIMA$(0, 1, 1)$ model (which has constant parameters and appears to be more of global type) and the random walk plus noise model (which allows the local mean level to change through time and so is more of local type).

In fact, there are similarities in that both models are non-stationary but neither implies that the trend is a deterministic function of time. This does illustrate that there may be no clearcut distinction between global and local models. Even so, the analyst may be wise to lean towards local-type models.

Simple or complicated? The model-building *Principle of Parsimony* (or Occam's Razor – see Section 3.5.4) says that the smallest possible number of parameters should be used so as to give an adequate representation of the given data. The more complicated the model, the more possibilities there will be for departures from model assumptions. The dangers of overfitting are particularly acute when constructing multiple regression, autoregressive and neural network models, as illustrated in the latter case by Example 8.4. Unfortunately, these dangers are not always heeded.

The inexperienced analyst may intuitively expect more complicated models to give better forecasts, but this is usually not the case. A more complicated model may reduce bias (though not if unnecessary terms are included), but may also increase variance, because more parameters have to be estimated (Breiman, 1992, p. 738). For example, Davies and Newbold (1980) show that, although an MA(1) model can be approximated arbitrarily closely by an AR model of high order, the effect of having to estimate additional parameters from finite samples is that forecast error variance gets worse for higher order models. Empirical evidence (e.g. Fildes and Makridakis, 1995) also points towards simpler models, and it is sad that empirical findings are often ignored by theoreticians, even though the Optimism Principle, introduced in Section 8.4, provides an explanation as to why complicated models may give better fits but worse predictions.

Is the method robust? There is much to be said for choosing a forecasting method, not because it is optimal for a particular model, but rather because it can adapt to changes in the underlying model structure and because it gives good forecasts in practice. Exponential smoothing is an excellent example of a robust forecasting procedure, because it is optimal for several different types of underlying model. For example, the empirical results of Chen (1997) suggest that the seasonal version of exponential smoothing (Holt-Winters) is more robust to departures from the implied underlying model than alternative methods based on seasonal ARIMA, structural and regression models.

When a model is finally chosen for a given set of data, it is always sensible to ask if the model is robust to changes in the model assumptions. One way of doing this is to carry out some sort of *sensitivity analysis*. This involves making small changes to the model assumptions in order to see how stable the deductions (including forecasts) from the model are. Sensitivity analysis is well established in some areas of science and engineering (e.g. Saltelli and Scott, 1997), but has not yet received the attention it deserves in time-series analysis.

a genuine hold-out sample. It is a sad fact that many analysts overlook
this point and regard the forecast accuracy obtained in this way as being
a genuine out-of-sample result. However, the Optimism Principle applies
here as well. The real forecast accuracy will generally be worse than that
found for the 'holdout' sample which has already been used to help select
the 'best' model.

8.5.5 Handling structural breaks

The most difficult type of model uncertainty to cope with is that associated
with a sudden change to the underlying model leading to what is variously
called a *structural break* or *change point* or *regime switch*. This may reveal
itself as a sudden jump in level, a sudden change in gradient (e.g. from
positive to negative) or some other obvious change in structure. An example
of the latter is a change from a regular seasonal pattern to a completely
different seasonal pattern, or to no discernible seasonal pattern at all,
perhaps as a result of a change in government policy. Figure 8.1 shows
a time series where the seasonal effect appears to disappear for about
two years and then re-appear. This makes life particularly difficult for the
analyst!

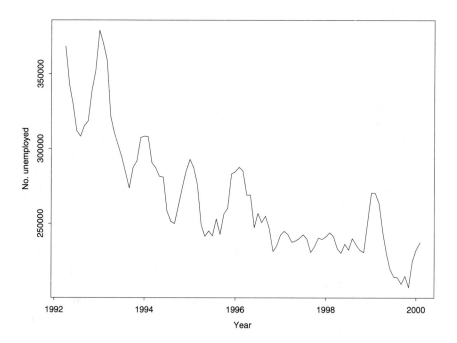

Figure 8.1 *The time plot, from April 1992 to February 2000, of monthly
ILO unemployment in the U.K. for males aged 25–49 who have been
unemployed for less than 6 months.*

History shows that structural breaks *do* occur and may ruin forecasts. Indeed, structural breaks will sometimes be revealed precisely because a forecast has failed, perhaps because a realized outcome is outside its prediction interval or because the forecasting system systematically over- or under-forecasts over several consecutive periods.

Thus there are (at least) two separate issues here. (1) How can we detect a structural break in past data and how should we respond when such a break does occur? (2) Is it possible to anticipate a structural break that may occur some time in the future, and how should forecasts be made when such an eventuality is thought to be possible?

Starting with the second issue, it is well established that sudden changes are difficult to forecast (e.g. Makridakis, 1988). This is particularly true when they are caused by an unforeseen event, such as the Gulf war, but even when they are caused by a known external event such as an expected change in government fiscal policy. Of course, univariate methods cannot be expected to forecast structural breaks as there are no explanatory variables to act as leading indicators. For multivariate modelling, there is more chance of anticipating sudden change, although this will still depend on how the break affects the relationship between variables. When a break occurs, it may, or may not, manifest itself in such a way that the change is related across variables in an analogous way to cointegration. If not, then predictive failure is more likely to occur. In particular, a model that imposes cointegration may fail to 'error-correct' following a structural break and this will cause problems (Clements and Hendry, 1998b, p. 213). Similar considerations may help to explain the poor performance of those econometric models, which are not robust to structural breaks. The ideal type of predictor variable is one that is causally related to the output and also cointegrated with it through sudden changes, so that it can continue to be useful for forecasting in the presence of a structural break.

Returning to the first issue, it may seem easier to detect sudden changes in past data, perhaps simply by looking at the time plot as in Figure 8.1. However, sudden changes may only become clear retrospectively when quite a few observations have been taken *after* the break. Thus, in Figure 8.1, the sudden disappearance of the seasonal effect in 1996/97 would only become clear when at least one year's data is available after the change. The same applies when the seasonal effect returns. This break was probably caused by a government decision to change the way that unemployment benefits were paid and this contextual knowledge might have led to an earlier diagnosis of the first change. However, the return of the seasonal effect in 1999 was completely unexpected and still unexplained at the time of writing. Some tests have been devised for assessing structural breaks, but they need careful specification (e.g. is the test for a permanent or temporary change in level), and may need many observations after the change. Thus they are primarily intended for retrospective analysis, rather than for an immediate assessment of a new unusual observation.

When a structural break has been found in past data, this should be taken into account when computing forecasts. This can be done in various ways, such as incorporating an intervention effect into an ARIMA model (see Section 5.6), applying a suitable multivariate or judgemental approach, or using a method that is robust to sudden changes. As regards the last possibility, one useful general effect, that is relevant to handling structural breaks, is that differencing can mitigate the effects of a jump in level (Clements and Hendry, 1998b). Although a model fitted to all the differenced data will not be exactly true (especially at the break point), the use of a model for differenced data means that it will adapt quickly away from the actual break point. Thus, in some sense, the model is robust to sudden changes. The alternative of using undifferenced data and fitting a different intercept before and after the change, will be less robust and harder to implement.

Although univariate models are generally unable to predict structural breaks, it is worth adding that such models do have other attractive properties. In particular, a simple model may be more robust to changes in the underlying model than more complicated alternatives. On the other hand, a multivariate model will be needed if the analyst wants to understand the reasons for a structural break, as will be necessary if policy analysis is to be attempted as well as forecasting.

8.6 Summary and discussion

The forecasting literature relies heavily on classical theory which generally assumes that the model for a given set of data is known and pre-specified. However, this bears little resemblance to the situation typically faced by forecasters in practice. A model for time-series data is usually formulated from the same data used to produce forecasts. Many models may be entertained and a single model is usually selected as the 'winner' even when other models give nearly as good a fit. Given that the wrong model may be selected or that a true model may not exist anyway, it follows that *model uncertainty* will affect most real statistical problems. It is therefore unfortunate that the topic has received little attention from forecasters. The main message of this chapter is that, *when a time-series model is formulated and fitted to the same data, then inferences and forecasts made from the fitted model will be biased and seriously over-optimistic when the prior data-dependent model-selection process is ignored.*

In more detail, the following general points can be made:

(a) Standard least-squares theory does not apply when the same data are used to formulate and fit a model. Estimates of the model parameters, including the residual variance, are likely to be biased.

(b) Models with more parameters may give a better fit, but worse out-of-sample predictions. Thus, when comparing the fit of different models, a measure of fit must be used which penalizes the introduction of additional parameters. The AIC_C or BIC is recommended in preference to the ordinary AIC for choosing between time-series forecasting models.

(c) The analyst typically thinks the fit is better than it really is (the Optimism Principle), and prediction intervals are generally too narrow, partly because residual variance tends to be underestimated and partly because prediction intervals fail to take full account of model uncertainty.

(d) The frequentist approach does not adapt naturally to cope with model uncertainty, though some progress can be made with resampling and other computational methods. Bayesian model averaging offers a promising alternative approach, even to analysts who are not Bayesian. However, difficulties arise whichever approach is adopted, and there appears to be no simple general theoretical 'fix'.

So how should the results in this chapter affect the way that a forecaster proceeds? Despite the difficulties in giving general advice, the following guidelines will be given to round off this book. First, the following well-established guidelines should always be followed:

(i) Clarify the objectives of the forecasting exercise.

(ii) Find out exactly how a forecast will actually be used.

(iii) Find out if a model is required for descriptive purposes as well as for prediction.

(iv) Ask questions to get background information as to a suitable class of models.

(v) Carry out a preliminary examination of the data to check for errors, outliers and missing observations as well as to see if the data have important features such as trend, seasonality, turning points and structural breaks. Incorporate the latter features into any model that is built.

The following additional guidelines arise from the material in this chapter, namely:

(vi) Be alert to the insidious existence of model uncertainty but be aware that there is no simple, general way of overcoming the problem.

(vii) Realise that the computer-based revolution in time-series modelling means that the analyst typically looks at a (very) large number of models. This may lead to biases when inference and forecasting follows data-based model selection.

(viii) Realise that more complicated models, while often appearing to give a better fit, do not necessarily produce better out-of-sample forecasts. Select a model on the basis of within-sample fit by using the AIC_C or BIC, so as to adequately penalize the addition of extra parameters. Realise that forecasts from a best-fitting model will generally not be as good as expected and that prediction intervals will generally be too narrow. The use of data splitting to check out-of-sample forecasting ability should avoid being fooled by optimistic within-sample results.

(ix) Consider the following alternatives to the use of a single best-fitting model; (i) use different models for different purposes; (ii) use scenario analysis to get a range of forecasts based on different assumptions about the future; (iii) use more than one model by some sort of mixing, such as taking a weighted combination of forecasts or using Bayesian model averaging; (iv) use a forecasting method which is not model based but which is designed to be robust to outliers and other peculiarities, as well as to changes in the underlying model.

References

The numbers in the square brackets at the end of each reference are the section numbers in which the reference is cited. There may be more than one citation in a section.

Abraham, B. and Ledolter, J. (1983) *Statistical Methods for Forecasting.* New York: Wiley. [1.5, 7.1, 7.5]

Abraham, B. and Ledolter, J. (1986) Forecast functions implied by autoregressive integrated moving average models and other related forecast procedures. *Int. Stat. Review,* **54**, 51-66. [4.3, 4.4]

Anderson, T. W. (1971) *The Statistical Analysis of Time Series.* New York: Wiley. [1.5]

Andrews, R. L. (1994) Forecasting performance of structural time series models. *J. Bus. Econ. Statist.,* **12**, 129-133. [6.4]

Ansley, C. F. and Kohn, R. (1986) Prediction mean squared error for state space models with estimated parameters. *Biometrika,* **73**, 467-473. [7.4]

Ansley, C. F. and Newbold, P. (1980) Finite sample properties of estimators for autoregressive-moving average models. *J. Econometrics,* **13**, 159-183. [4.2, 7.4]

Ansley, C. F. and Newbold, P. (1981) On the bias in estimates of forecast mean squared error. *J. Amer. Statist. Assoc.,* **76**, 569-578. [7.4]

Aoki, M. (1987) *State Space Modeling of Time Series.* Berlin: Springer-Verlag. [3.2]

Armstrong, J. S. (1970) How to avoid exploratory research. *J. Advertising Research,* **10**, 27-30. [5.2]

Armstrong, J. S. (1985) *Long-Range Forecasting,* 2nd edn. New York: Wiley. [5.2, 7.5, 8.5]

Armstrong, J. S. (2001) *Principles of Forecasting: A Handbook for Researchers and Practitioners.* Norwell, MA: Kluwer. To appear. [1.5]

Armstrong, J. S. and Collopy, F. (1992) Error measures for generalizing about forecasting methods: Empirical comparisons. *Int. J. Forecasting,* **8**, 69-80. [6.3]

Armstrong, J. S. and Collopy, F. (1998) Integration of statistical methods and judgment for time series forecasting: Principles from empirical research. In *Forecasting with Judgment* (eds. G. Wright and P. Goodwin), pp. 269-293. New York: Wiley. [1.1]

Armstrong, J. S. and Collopy, F. (2001) Identification of asymmetric prediction intervals through causal forces. *J. Forecasting.* To appear. [7.7]

Armstrong, J. S. and Lusk, E. J. (eds.) (1983) Commentary on the Makridakis time series competition (M-competition). *J. of Forecasting*, **2**, 259-311. [6.4]

Ashley, R. (1988) On the relative worth of recent macroeconomic forecasts. *Int. J. Forecasting*, **4**, 363-376. [5.1, 7.7]

Baillie, R. T. (1979) Asymptotic prediction mean squared error for vector autoregressive models. *Biometrika*, **66**, 675-678. [7.4]

Ball, M. and Wood, A. (1996) Trend growth in post-1850 British economic history: the Kalman filter and historical judgment. *The Statistician*, **45**, 143-152. [2.3]

Banerjee, A., Dolado, J., Galbraith, J. W. and Hendry, D. F. (1993) *Co-Integration, Error-Correction, and the Econometric Analysis of Non-Stationary Data*. Oxford: Oxford Univ. Press. [5.4]

Barnett, G., Kohn, R. and Sheather, S. (1996) Estimation of an autoregressive model using Markov chain Monte Carlo. *J. Econometrics*, **74**, 237-254. [8.5]

Bates, J. M. and Granger, C. W. J. (1969) The combination of forecasts. *Op. Res. Q.*, **20**, 451-468. [4.3]

Bell, W. R. and Hillmer, S. C. (1983) Modelling time series with calendar variation. *J. Amer. Statist. Assoc.*, **78**, 526-534. [2.3]

Beran, J. (1992) Statistical methods for data with long-range dependence. *Statist. Sci.*, **7**, 404-427. [3.1]

Beran, J. (1994) *Statistics for Long-Memory Processes*. New York: Chapman and Hall. [3.1]

Berliner, L. M. (1991) Likelihood and Bayesian prediction of chaotic systems. *J. Amer. Statist. Assoc.*, **86**, 938-952. [3.4]

Bernado, J. M. and Smith, A. F. M. (1994) *Bayesian Theory*. Chichester: Wiley. [8.2]

Bianchi, C., Calzolari, G. and Brillet, J-L. (1987) Measuring forecast uncertainty. *Int. J. Forecasting*, **3**, 211-227. [7.5]

Bishop, C. M. (1995) *Neural Networks for Pattern Recognition*. Oxford: Clarendon Press. [3.4]

Boero, G. (1990) Comparing *ex-ante* forecasts from a SEM and VAR model: An application to the Italian economy. *J. Forecasting*, **9**, 13-24. [6.4]

Bollerslev, T., Chou, Y. and Kroner, K. F. (1992) ARCH models in finance. *J. of Econometrics*, **52**, 5-59. [3.4]

Bollerslev, T., Engle, R. F. and Nelson, D. B. (1994) ARCH models. In *Handbook of Econometrics, Vol. IV* (eds. R. F. Engle, and D. L. McFadden), pp. 2959-3038. Amsterdam: Elsevier. [3.4]

Bowerman, B. L. and Koehler, A. B. (1989) The appropriateness of Gardner's simple approach and Chebychev prediction intervals. Paper presented at the 9th International Symposium on Forecasting in Vancouver. [7.5, 7.7]

Bowerman, B. L. and O'Connell, R. T. (1987) *Time Series Forecasting*, 2nd edn. Boston: Duxbury Press. [1.5, 7.1, 7.5]

Box, G. E. P. and Jenkins, G. M. (1970) *Time-Series Analysis, Forecasting and Control*. San Francisco: Holden-Day (revised edn., 1976). [1.5, 3.1]

Box, G. E. P., Jenkins, G. M. and Reinsel, G. C. (1994) *Time Series Analysis,*

Forecasting and Control, 3rd edn. Englewood Cliffs, NJ: Prentice-Hall. [1.5, 2.3, 3.1, 3.5, 4.2, 5.1–5.3, 5.6, 7.4, 7.5, 7.7, 7.8, 8.2]

Box, G. E. P. and Newbold, P. (1971) Some comments on a paper of Coen, Gomme and Kendall. *J. Roy. Statist. Soc. A*, **134**, 229-240. [5.1, 5.2]

Breiman, L. (1992) The little bootstrap and other methods for dimensionality selection in regression: X-fixed prediction error. *J. Amer. Statist. Assoc.*, **87**, 738-754. [8.5]

Brillinger, D. R. (1981) *Time Series: Data Analysis and Theory*, expanded edn. New York: Holt, Rinehart and Winston. [1.5]

Brock, W. A. and Potter, S. M. (1993) Non-linear time series and Macroeconometrics. In *Handbook of Statistics, Volume 11, Econometrics* (eds. G. S. Maddala, C. R. Rao and H. D. Vinod), pp. 195-229. Amsterdam: North-Holland. [3.4]

Brockwell, P. J. and Davis, R. A. (1991) *Time Series: Theory and Methods*, 2nd edn. New York: Springer-Verlag. [1.5, 3.1, 7.1]

Brockwell, P. J. and Davis, R. A. (1996) *Introduction to Time Series and Forecasting*. New York: Springer-Verlag. [1.5, 3.1, 4.3, 6.4]

Brown, R. G. (1963) *Smoothing, Forecasting and Prediction*. Englewood Cliffs, NJ: Prentice-Hall. [4.3, 7.5]

Brown, R. G. (1967) *Decision Rules for Inventory Management*. New York: Holt, Rinehart and Winston. [7.5]

Burnham, K. P. and Anderson, D. R. (1998) *Model Selection and Inference*. New York: Springer-Verlag. [3.5, 8.3, 8.4]

Butler, R. and Rothman, E. D. (1980) Predictive intervals based on reuse of the sample. *J. Amer. Statist. Assoc.*, **75**, 881-889. [7.5]

Butter, F. A. G. den and Fase, M. M. G. (1991) *Seasonal Adjustment as a Practical Problem*. Amsterdam: Elsevier. [2.3]

Callen, J. L., Kwan, C. C. Y., Yip, P. C. Y. and Yuan, Y. (1996) Neural network forecasting of quarterly accounting earnings. *Int. J. Forecasting*, **12**, 475-482. [6.4]

Carbone, R. and Armstrong, J. S. (1982) Evaluation of extrapolative forecasting methods: results of a survey of academicians and practitioners. *J. of Forecasting*, **1**, 215-217. [6.4]

Carlstein, E. (1992) Resampling techniques for stationary time series: Some recent developments. In *New Directions in Time Series Analysis, Part I*, (eds. D. Brillinger et al.), pp. 75-85. New York: Springer-Verlag. [7.5]

Chappell, D. et al. (1996) A threshold model for the French Franc/Deutschmark exchange rate. *J. Forecasting*, **15**, 155-164. [3.4]

Chatfield, C. (1978) The Holt-Winters forecasting procedure. *Appl. Statist.*, **27**, 264-279. [6.5]

Chatfield, C. (1988a) What is the best method of forecasting? *J. Appl. Statist.*, **15**, 19-38. [6.0, 6.4, 6.5]

Chatfield, C. (1988b) Apples, oranges and mean square error. *Int. J. of Forecasting*, **4**, 515-518. [6.3, 6.4]

Chatfield, C. (1993) Calculating interval forecasts (with discussion). *J. Bus. Econ. Statist.*, **11**, 121-144. [7.1]

Chatfield, C. (1995a) *Problem-Solving: A Statistician's Guide*, 2nd edn. London: Chapman and Hall. [2.3, 3.5]

Chatfield, C. (1995b) Model uncertainty, data mining and statistical inference (with discussion). *J. Roy. Statist. Soc. A*, **158**, 419-466. [3.5, 8.2–8.5]

Chatfield, C. (1995c) Positive or negative? *Int. J. of Forecasting*, **11**, 501-502. [6.4]

Chatfield, C. (1996a) *The Analysis of Time Series*, 5th edn. London: Chapman and Hall. [1.5, 2.3, 3.2, 3.4, 4.2, 4.3, 5.2, 6.0, 6.5, 7.1]

Chatfield, C. (1996b) Model uncertainty and forecast accuracy. *J. Forecasting*, **15**, 495-508. [7.1, 7.7, 8.3]

Chatfield, C. (1997) Forecasting in the 1990s. *The Statistician*, **46**, 461-473. [1.5, 6.0]

Chatfield, C. (2001) Prediction intervals. In *Principles of Forecasting: A Handbook for Researchers and Practitioners* (ed. J. S. Armstrong), Norwell, MA: Kluwer. [7.1]

Chatfield, C. and Faraway, J. (1996) Forecasting sales data with neural nets: A case study (in French). *Recherche et Applications en Marketing*, **11**, 29-41. [8.3]

Chatfield, C. and Koehler, A. B. (1991) On confusing lead time demand with *h*-period-ahead forecasts. *Int. J. Forecasting*, **7**, 239-240. [7.5]

Chatfield, C., Koehler, A. B., Ord, J. K. and Snyder, R. D. (2001) Models for exponential smoothing: A review of recent developments. *The Statistician*, **50**. To appear. [4.3]

Chatfield, C. and Prothero, D. L. (1973) Box-Jenkins seasonal forecasting: Problems in a case study (with discussion). *J. Roy. Statist. Soc. A*, **136**, 295-352. [7.5]

Chatfield, C. and Yar, M. (1988) Holt-Winters forecasting: Some practical issues. *The Statistician*, **37**, 129-140. [4.3, 6.5]

Chatfield, C. and Yar, M. (1991) Prediction intervals for multiplicative Holt-Winters. *Int. J. Forecasting*, **7**, 31-37. [7.5, 7.7]

Chen, C. and Liu, L-M. (1993) Forecasting time series with outliers. *J. Forecasting*, **12**, 13-35. [2.3]

Chen, C. (1997) Robustness properties of some forecasting methods for seasonal time series: A Monte Carlo study. *Int. J. Forecasting*, **13**, 269-280. [6.4, 8.5]

Choi, B. (1992) *ARMA Model Identification*. New York: Springer-Verlag. [3.5]

Choudhury, A. H., Hubata, R. and Louis, R. D. St. (1999) Understanding time-series regression estimators. *Amer. Statist.*, **53**, 342-348. [5.2]

Christofferson, P. F. (1998) Evaluating interval forecasts. *Int. Economic Review*, **39**, 841-862. [7.7]

Christoffersen, P. F. and Diebold, F. X. (1998) Cointegration and long-horizon forecasting. *J. Bus. Econ. Statist.*, **16**, 450-8. [5.4, 6.3]

Church, K. B. and Curram, S. P. (1996) Forecasting consumers' expenditure: A comparison between econometric and neural network models. *Int. J. Forecasting*, **12**, 255-267. [6.4]

Clemen, R. T. (1989) Combining forecasts: A review and annotated bibliography. *Int. J. Forecasting*, **5**, 559-583. [4.3, 7.5, 8.5]

Clements, M. P. and Hendry, D. F. (1996) Forecasting in macro-economics. In *Time Series Models* (eds. D. R. Cox, D. V. Hinkley and O. E. Barndorff-Nielsen), pp. 101-141. London: Chapman and Hall. [5.5]

Clements, M. P. and Hendry, D. F. (1998a) *Forecasting Economic Time Series*. Cambridge: Cambridge Univ. Press. [4.3, 5.2, 5.4, 6.3, 8.1, 8.4, 8.5]

Clements, M. P. and Hendry, D. F. (1998b) Forecasting economic processes (with discussion). *Int. J. Forecasting*, **14**, 111-143. [7.7, 8.5]

Clements, M. P. and Krolzig, H-M. (1998) A comparison of the forecast performance of Markov-switching and threshold autoregressive models of U.S. GNP. *Econometrics J.*, **1**, C47-C75. [3.4, 6.4]

Clements, M. P. and Smith, J. (1997) The performance of alternative forecasting methods for SETAR models. *Int. J. Forecasting*, **13**, 463-475. [3.4, 4.2, 6.4]

Collins, S. (1991) Prediction techniques for Box-Cox regression models. *J. Bus. Econ. Statist.*, **9**, 267-277. [7.5]

Collopy, F. and Armstrong, J. S. (1992) Rule-based forecasting: Development and validation of an expert systems approach to combining time series extrapolations. *Man. Sci.*, **38**, 1394-1414. [6.5]

Copas, J. B. (1983) Regression, prediction and shrinkage (with discussion). *J. Roy. Statist. Soc. B*, **45**, 311-354. [8.4]

Corker, R. J., Holly, S. and Ellis, R. G. (1986) Uncertainty and forecast precision. *Int. J. Forecasting*, **2**, 53-69. [7.5]

Crato, N. and Ray, B. K. (1996) Model selection and forecasting for long-range dependent processes. *J. Forecasting*, **15**, 107-125. [3.1, 6.4]

Dacco, R. and Satchell, S. (1999) Why do regime-switching models forecast so badly? *J. Forecasting*, **18**, 1-16. [3.4]

Dalrymple, D. J. (1987) Sales forecasting practices: Results from a United States survey. *Int. J. Forecasting*, **3**, 379-391. [7.1]

Dangerfield, B. J. and Morris, J. S. (1992) Top-down or bottom-up: Aggregate versus disaggregate extrapolations. *Int. J. Forecasting*, **8**, 233-241. [2.1]

Davies, N. and Newbold, P. (1980) Forecasting with misspecified models. *Applied Statistics*, **29**, 87-92. [8.5]

DeJong, D. N. and Whiteman, C. H. (1993) Unit roots in U.S. macroeconomic time series: A survey of classical and Bayesian perspectives. In *New Directions in Time Series Analysis: Part II* (eds. D. Brillinger et al.), pp. 43-59. New York: Springer-Verlag. [3.1]

Dhrymes, P. (1997) *Time Series, Unit Roots and Cointegration*. San Diego: Academic Press. [5.4]

Diebold, F. X. (1998) *Elements of Forecasting*. Cincinnati: South-Western College Publishing. [1.5, 2.3, 3.1, 4.3]

Diebold, F. X. and Kilian, L. (2000) Unit root tests are useful for selecting forecasting models. *J. Bus. Econ. Statist.*, **18**, 265-273. [3.1]

Diggle, P. J. (1990) *Time Series: A Biostatistical Introduction*. Oxford: Oxford Univ. Press. [1.5]

Draper, D. (1995) Assessment and propagation of model uncertainty (with discussion). *J. Roy. Statist. Soc. B*, **57**, 45-97. [7.5, 8.4, 8.5]

Enders, W. (1995) *Applied Econometric Time Series*. New York: Wiley. [1.5, 3.1, 3.4]

Engle, R. F. and Granger, C. W. J. (1991) *Long-Run Economic Relationships: Readings in Cointegration*. Oxford: Oxford Univ. Press. [5.4]

Engle, R. F. and Yoo, B. S. (1987) Forecasting and testing in co-integrated systems. *J. Econometrics*, **35**, 143-159. [7.8]

Fair, R. C. (1980) Estimating the expected predictive accuracy of econometric models. *Int. Econ. Review*, **21**, 355-378. [7.5]

Faraway, J. (1992) On the cost of data analysis. *J. Computational and Graphical Statistics*, **1**, 213-229. [8.4, 8.5]

Faraway, J. (1998) Data splitting strategies for reducing the effect of model selection on inference. *Comp. Science and Stats.*, **30**, 332-341. [8.5]

Faraway, J. and Chatfield, C. (1998) Time series forecasting with neural networks: A comparative study using the airline data. *Appl. Statist.*, **47**, 231-250. [3.4, 3.5, 6.4, 8.3]

Farnum, N. R. and Stanton, L. W. (1989) *Quantitative Forecasting Methods*. Boston: PWS-Kent. [5.2]

Fildes, R. (1983) An evaluation of Bayesian forecasting. *J. Forecasting*, **2**, 137-150. [6.4]

Fildes, R. (1985) Quantitative forecasting: The state of the art. Econometric models. *J. Op. Res. Soc.*, **36**, 549-580. [6.4]

Fildes, R. (1992) The evaluation of extrapolative forecasting methods. *Int. J. Forecasting*, **8**, 81-98. [6.3, 6.5]

Fildes, R. and Makridakis, S. (1995) The impact of empirical accuracy studies on time series analysis and forecasting. *Int. Statist. Review*, **63**, 289-308. [6.4, 7.7, 8.3, 8.5]

Findley, D. F. (1986) On bootstrap estimates of forecast mean square errors for autoregressive processes, *In Computer Science and Statistics: 17th Symposium on the Interface* (ed. D.M. Allen), pp. 11-17. Amsterdam: North-Holland. [7.4, 7.5]

Findley, D. F., Monsell, B. C., Bell, W. R., Otto, M. C. and Chen, B-C. (1998) New capabilities and methods of the X-12-ARIMA seasonal adjustment program (with discussion and reply). *J. Bus. Econ. Statist.*, **16**, 127-177. [2.3]

Franses, P. H. (1996) *Periodicity and Stochastic Trends in Economic Time Series*. Oxford: Oxford Univ. Press. [3.1]

Franses, P. H. (1998) *Time Series Models for Business and Economic Forecasting*. Cambridge: Cambridge Univ. Press. [1.5, 3.4]

Franses, P. H. and Kleibergen, F. (1996) Unit roots in the Nelson-Plosser data: Do they matter for forecasting? *Int. J. Forecasting*, **12**, 283-288. [2.3, 3.1]

Franses, P. H. and Ooms, M. (1997) A periodic long-memory model for quarterly UK inflation. *Int. J. Forecasting*, **13**, 117-126. [6.4]

Freedman, D. A. and Peters, S. C. (1984a) Bootstrapping a regression equation: Some empirical results. *J. Amer. Statist. Assoc.*, **79**, 97-106. [7.5]

Freedman, D. A. and Peters, S. C. (1984b) Bootstrapping an econometric model: Some empirical results. *J. Bus. Econ. Statist.*, **2**, 150-158. [7.5]

Fuller, W. A. (1996) *Introduction to Statistical Time Series*, 2nd edn. New York: Wiley. [1.5]

Fuller, W. A. and Hasza, D. P. (1981) Properties of predictors for autoregressive time series. *J. Amer. Statist. Assoc.*, **76**, 155-161. [7.4]

Gardner, E. S. Jr. (1983) Automatic monitoring of forecast errors. *J. Forecasting*, **2**, 1-21. [3.5, 7.7]

Gardner, E. S. Jr. (1985) Exponential smoothing: The state of the art. *J. Forecasting*, **4**, 1-28. [4.3]

Gardner, E. S. Jr. (1988) A simple method of computing prediction intervals for time-series forecasts. *Man. Sci.*, **34**, 541-546. [7.7]

Gardner, E. S. Jr. and McKenzie, E. (1985) Forecasting trends in time series. *Man. Sci.*, **31**, 1237-46. [4.3]

Geisser, S. (1993) *Predictive Inference: An Introduction*. New York: Chapman and Hall. [8.5]

Gerlow, M. E., Irwin, S. H. and Liu, T-R. (1993) Economic evaluation of commodity price forecasting models. *Int. J. Forecasting*, **9**, 387-397. [6.3]

Gersch, W. and Kitagawa, G. (1983) The prediction of time series with trends and seasonalities. *J. Bus. Econ. Statist.*, **1**, 253-264. [8.5]

Ghysels, E., Granger, C. W. J. and Siklos, P. L. (1996) Is seasonal adjustment a linear or nonlinear data-filtering process? (with discussion). *J. Bus. Econ. Statist.*, **14**, 374-386. [2.3]

Gilchrist, W. (1976) *Statistical Forecasting*. London: Wiley. [7.5]

Gleick, J. (1987) *Chaos*. New York: Viking. [3.4]

Gomez, V. and Maravall, A. (2000) Seasonal adjustment and signal extraction in economic time series. In *A Course in Time Series Analysis* (eds. D. Pena, G. C. Tiao and R. S. Tsay), Chapter 8. New York: Wiley. [2.3]

Gooijer, J. G. de, Abraham, B., Gould, A. and Robinson, L. (1985). Methods for determining the order of an autoregressive-moving average process: A survey. *Int. Statist. Rev.*, **53**, 301-329. [3.5]

Gooijer, J. G. de and Franses, P. H. (1997) Forecasting and seasonality. *Int. J. Forecasting*, **13**, 303-305. [2.3]

Gooijer, J. G. de and Kumar, K. (1992) Some recent developments in non-linear time series modelling, testing and forecasting. *Int. J. Forecasting*, **8**, 135-156. [6.4]

Gourieroux, C. (1997) *ARCH Models and Financial Applications*. New York: Springer-Verlag. [3.4]

Granger, C. W. J. (1989) Combining forecasts – twenty years later. *J. Forecasting*, **8**, 167-173. [4.3]

Granger, C. W. J. (1992) Forecasting stock market prices: Lessons for forecasters. *Int. J. Forecasting*, **8**, 3-13. [3.4]

Granger, C. W. J. (1996) Can we improve the perceived quality of economic forecasts? *J. Appl. Econometrics*, **11**, 455-473. [7.7]

Granger, C. W. J. (1999) *Empirical Modeling in Economics.* Cambridge: Cambridge Univ. Press. [3.5]

Granger, C. W. J. and Andersen, A. P. (1978) *An Introduction to Bilinear Time Series Models.* Göttingen: Vandenhoeck and Ruprecht. [3.4]

Granger, C. W. J. and Joyeux, R. (1980) An introduction to long-range time series models and fractional differencing. *J. Time Series Analysis*, **1**, 15-30. [3.1]

Granger, C. W. J., King, M. L. and White, H. (1995) Comments on testing economic theories and the use of model selection criteria. *J. Econometrics*, **67**, 173-187. [3.5]

Granger, C. W. J. and Newbold, P. (1974) Spurious regressions in econometrics. *J. Econometrics*, **2**, 111-120. [5.2]

Granger, C. W. J. and Newbold, P. (1986) *Forecasting Economic Time Series*, 2nd edn. New York: Academic Press. [1.5, 3.4, 4.3, 5.1, 5.2, 5.5, 6.2, 7.1, 7.4, 7.5]

Granger, C. W. J. and Teräsvirta, T. (1993) *Modelling Nonlinear Economic Relationships.* New York: Oxford Univ. Press. [5.6]

Grigoletto, M. (1998) Bootstrap prediction intervals for autoregressions: some alternatives. *Int. J. Forecasting*, **14**, 447-456. [7.5]

Groot, C. de and Würtz, D. (1991) Analysis of univariate time series with connectionist nets: A case study of two classical examples. *Neurocomputing*, **3**, 177-192. [3.4]

Hahn, G. J. and Meeker, W. Q. (1991) *Statistical Intervals: A Guide for Practitioners.* New York: Wiley. [7.1]

Hamilton, J. D. (1994) *Time Series Analysis.* Princeton, NJ: Princeton Univ. Press. [1.5, 3.1, 3.4, 4.1, 5.2, 5.4]

Harrison, P. J. (1965) Short-term sales forecasting. *Appl. Statist.*, **14**, 102-139. [4.3]

Harrison, P. J. (1967) Exponential smoothing and short-term sales forecasting. *Man. Sci.*, **13**, 821-842. [7.5]

Harvey, A. C. (1989) *Forecasting, Structural Time Series Models and the Kalman Filter.* Cambridge: Cambridge Univ. Press. [1.5, 3.2, 5.6, 6.4, 7.1, 7.5]

Harvey, A. C. (1990) *The Econometric Analysis of Time Series*, 2nd edn. Hemel Hempstead, U.K.: Philip Allan. [1.5, 7.5]

Harvey, A. C. (1993) *Time Series Models*, 2nd edn. New York: Harvester Wheatsheaf. [1.5, 3.1, 3.4, 4.2, 4.4]

Haslett, J. (1997) On the sample variogram and the sample autocovariance for non-stationary time series. *The Statistician*, **46**, 475-485. [3.5]

Haywood, J. and Tunnicliffe Wilson, G. (1997) Fitting time series models by minimizing multistep-ahead errors: a frequency domain approach. *J. Roy. Statist. Soc. B*, **59**, 237-254. [8.5]

Hibon, M. and Makridakis, S. (1999) The M3-Competition. Paper presented to the 19th Int. Symposium on Forecasting, Washington, D.C. [6.4]

Hill, T., Marquez, L., O'Connor, M. and Remus, W. (1994) Artificial neural network models for forecasting and decision making. *Int. J. Forecasting*, **10**, 5-15. [6.4]

Hill, T., O'Connor, M. and Remus, W. (1996) Neural network models for time series forecasts. *Man. Sci.*, **42**, 1082-1092. [6.4]

Hjorth, U. (1987) On model selection in the computer age. Technical report no. LiTH-MAT-R-87-08, Linköping University, Sweden. [8.3, 8.4]

Hjorth, U. (1994) *Computer Intensive Statistical Methods - Validation Model Selection and Bootstrap*. London: Chapman and Hall. [7.5, 8.3, 8.4]

Hoptroff, R. (1993) The principles and practice of time series forecasting and business modelling using neural nets. *Neur. Comput. Applic.*, **1**, 59-66. [6.4]

Hylleberg, S. (ed.) (1992) *Modelling Seasonality*. Oxford: Oxford Univ. Press. [2.3]

Hyndman, R. J. (1995) Highest density forecast regions for non-linear and non-normal time series models. *J. Forecasting*, **14**, 431-441. [7.1, 7.5]

Isham, V. (1993) Statistical aspects of chaos. In *Networks and Chaos - Statistical and Probabilistic Aspects* (eds. O. E. Barndorff-Nielsen et al.), pp. 124-200. London: Chapman and Hall. [3.4]

Janacek, G. and Swift, L. (1993) *Time Series: Forecasting, Simulation, Applications*. Chichester, U.K.: Ellis Horwood. [1.5, 3.2, 4.2]

Jenkins, G. M. (1974) Contribution to the discussion of Newbold and Granger (1974). *J. Roy. Statist. Soc A*, **137**, 148-150. [4.4]

Jenkins, G. M. (1979) *Practical Experiences with Modelling and Forecasting Time Series*. Jersey: Gwilym Jenkins and Partners (Overseas) Ltd. [5.2, 6.4]

Jenkins, G. M. and McLeod, G. (1982) *Case Studies in Time Series Analysis*, Vol. 1. Lancaster: Gwilym Jenkins and Partners Ltd. [6.4]

Johansen, S. (1996) Likelihood-based inference for cointegration of some nonstationary time series. In *Time Series Models* (eds. D. R. Cox, D. V. Hinkley and O. E. Barndorff-Nielsen), pp. 71-100. London: Chapman and Hall. [5.4]

Johnston, F. R. and Harrison, P. J. (1986) The variance of lead-time demand. *J. Op. Res. Soc.*, **37**, 303-308. [7.5]

Judge, G. G. and Bock, M. E. (1978) *The Statistical Implications of Pre-test and Stein-rule Estimators in Econometrics*. Amsterdam: North Holland. [8.3]

Kadiyala, K. R. and Karlsson, S. (1993) Forecasting with generalized Bayesian vector autoregressions. *J. Forecasting*, **12**, 365-378. [5.3, 6.4]

Kantz, H. and Schreiber, T. (1997) *Nonlinear Time Series Analysis*. Cambridge: Cambridge Univ. Press. [3.4]

Kass, R. E. and Raftery, A. E. (1995) Bayes factors. *J. Amer. Statist. Assoc.*, **90**, 773-795. [8.5]

Kendall, M. G. and Ord, J. K. (1990) *Time Series*, 3rd edn. Sevenoaks, U.K.: Arnold. [1.5, 4.3, 7.4, 7.5]

Kendall, M. G., Stuart, A. and Ord, J. K. (1983) *The Advanced Theory of Statistics*, Vol. 3, 4th edn. London: Griffin. [1.5]

Kenny, P. B. and Durbin, J. (1982) Local trend estimation and seasonal adjustment of economic and social time series (with discussion). *J. Roy. Statist. Soc. A*, **145**, 1-41. [2.3]

Klein, L. R. (1988) The statistical approach to economics. *J. Econometrics*, **37**, 7-26. [5.5]

Koehler, A. B. (1990) An inappropriate prediction interval. *Int. J. Forecasting*, **6**, 557-558. [7.5]

Kulendran, N. and King, M. L. (1997) Forecasting international quarterly tourist flows using error-correction and time-series models. *Int. J. Forecasting*, **13**, 319-327. [5.4]

Lambert, P. and Lindsey, J. K. (1999) Analysing financial returns by using regression models based on non-symmetric stable distributions. *Appl. Statist.*, **48**, 409-424. [7.7]

Le, N. D., Raftery, A. E. and Martin, R. D. (1996) Robust Bayesian model selection for autoregressive processes with additive outliers. *J. Amer. Statist. Assoc.*, **91**, 123-131. [8.5]

Ledolter, J. (1989) The effect of additive outliers on the forecasts from ARIMA models. *Int. J. Forecasting*, **5**, 231-240. [2.3, 7.7]

Lefrancois, P. (1989) Confidence intervals for non-stationary forecast errors: Some empirical results for the series in the M-competition. *Int. J. Forecasting*, **5**, 553-557. [7.5]

Lefrancois, P. (1990) Reply to: Comments by C. Chatfield. *Int. J. Forecasting*, **6**, 561. [7.5]

Levenbach, H. and Cleary, J. P. (1984) *The Modern Forecaster*. Belmont, CA: Lifetime Learning Publications. [5.2]

Lin, J-L. and Granger, C. W. J. (1994) Forecasting from non-linear models in practice. *J. Forecasting*, **13**, 1-9. [4.2]

Lin, J-L. and Tsay, R. S. (1996) Co-integration constraint and forecasting: An empirical examination. *J. App. Econometrics*, **11**, 519-538. [5.4]

Luna, X. de (2000) Prediction intervals based on autoregression forecasts. *The Statistician*, **49**, 1-29. [7.4]

Lütkepohl, H. (1987) *Forecasting Aggregated Vector ARMA Processes*. Berlin: Springer-Verlag. [5.3]

Lütkepohl, H. (1993) *Introduction to Multiple Time Series Analysis*, 2nd edn. New York: Springer-Verlag. [5.3, 5.4, 5.7, 7.1, 7.4, 7.7, 7.8]

Makridakis, S. (1988) Metaforecasting. *Int. J. Forecasting*, **4**, 467-491. [7.7, 8.5]

Makridakis, S., Anderson, A., Carbone, R., Fildes, R., Hibon, M., Lewandowski, R., Newton, J., Parzen, E. and Winkler, R. (1982) The accuracy of extrapolation (time series) methods: Results of a forecasting competition. *J. Forecasting*, **1**, 111-153. [6.4]

Makridakis, S., Anderson, A., Carbone, R., Fildes, R., Hibon, M., Lewandowski, R., Newton, J., Parzen, E. and Winkler, R. (1984) *The Forecasting Accuracy of Major Time Series Methods*. New York: Wiley. [4.3, 6.4]

Makridakis, S., Chatfield, C., Hibon, M., Lawrence, M., Mills, T., Ord, K. and Simmons, L. F. (1993) The M2-competition: A real-time judgmentally based forecasting study (with commentary). *Int. J. Forecasting*, **9**, 5-29. [6.4]

Makridakis, S. and Hibon, M. (1979) Accuracy of forecasting: An empirical investigation (with discussion). *J. Roy. Statist. Soc. A*, **142**, 97-145. [6.4]

Makridakis, S. and Hibon, M. (1997) ARMA models and the Box-Jenkins methodology. *J. Forecasting*, **16**, 147-163. [4.2]

Makridakis, S., Hibon, M., Lusk, E. and Belhadjali, M. (1987) Confidence intervals: An empirical investigation of the series in the M-competition. *Int. J. Forecasting*, **3**, 489-508. [7.5, 7.7]

Makridakis, S. and Wheelwright, S. C. (1989) *Forecasting Methods for Management*, 5th edn. New York: Wiley. [5.2]

Makridakis, S., Wheelwright, S. C. and Hyndman, R. J. (1998) *Forecasting: Methods and Applications*, 3rd edn. New York: Wiley. [1.5]

Makridakis, S. and Winkler, R. L. (1989) Sampling distributions of post-sample forecasting errors. *Appl. Statist.*, **38**, 331-342. [7.5, 7.7]

Marriott, J. and Newbold, P. (1998) Bayesian comparison of ARIMA and stationary ARMA models. *Int. Statist. Rev.*, **66**, 323-336. [3.1]

Masarotto, G. (1990) Bootstrap prediction intervals for autoregressions. *Int. J. Forecasting*, **6**, 229-239. [7.5, 7.7]

McCullough, B. D. (1994) Bootstrapping forecast intervals: An application to AR(p) models. *J. Forecasting*, **13**, 51-66. [7.5]

McCullough, B. D. (1996) Consistent forecast intervals when the forecast-period exogenous variables are stochastic. *J. Forecasting*, **15**, 293-304. [7.5]

McCullough, B. D. (1998) Algorithm choice for (partial) autocorrelation functions. *J. of Economic and Social Measurement*, **24**, 265-278. [6.5]

McCullough, B. D. (2000) Is it safe to assume that software is accurate? *Int. J. Forecasting*, **16**, 349-357. [6.5]

McKenzie, E. (1986) Error analysis for Winters' additive seasonal forecasting system. *Int. J. Forecasting*, **2**, 373-382. [7.5]

McKenzie, M. D. (1999) Power transformation and forecasting the magnitude of exchange rate changes. *Int. J. Forecasting*, **15**, 49-55. [3.4]

McLain, J. O. (1988) Dominant tracking signals. *Int. J. Forecasting*, **4**, 563-572. [3.5]

McLeod, A. I. (1993) Parsimony, model adequacy and periodic correlation in time series forecasting. *Int. Statist. Rev.*, **61**, 387-393. [3.1, 6.4]

Meade, N. (1984) The use of growth curves in forecasting market development – a review and appraisal. *J. Forecasting*, **3**, 429-451. [3.3]

Meade, N. and Islam, T. (1995) Prediction intervals for growth curve forecasts. *J. Forecasting*, **14**, 413-430. [7.5]

Meade, N. and Smith, I. D. (1985) ARARMA vs. ARIMA – A study of the benefits of a new approach to forecasting. *Int. J. Management Sci.*, **13**, 519-534. [4.3]

Meese, R. and Geweke, J. (1984) A comparison of autoregressive univariate forecasting procedures for macroeconomic time series. *J. Bus. Econ. Statist.*, **2**, 191-200. [8.5]

Meese, R. and Rogoff, K. (1983) Empirical exchange rate models of the seventies: Do they fit out of sample? *J. Int. Economics*, **14**, 3-24. [6.4]

Miller, A. J. (1990) *Subset Selection in Regression*. London: Chapman and Hall. [7.5, 8.3, 8.4]

Mills, T. C. (1990) *Time Series Techniques for Economists*. Cambridge: Cambridge Univ. Press. [1.5]

Mills, T. C. (1999) *The Econometric Modelling of Financial Time Series*, 2nd edn. Cambridge: Cambridge Univ. Press. [1.5]

Montgomery, A. L., Zarnowitz, V., Tsay, R. S. and Tiao, G. C. (1998) Forecasting the U.S. unemployment rate. *J. Amer. Statist. Assoc.*, **93**, 478-493. [2.1, 3.4]

Montgomery, D. C., Johnson, L. A. and Gardiner, J. S. (1990) *Forecasting and Time Series Analysis*, 2nd edn. New York: McGraw-Hill. [1.5]

Murray, M. P. (1994) A drunk and her dog: An illustration of cointegration and error correction. *Amer. Statistician*, **48**, 37-9. [5.4]

Nelson, H. L. and Granger, C. W. J. (1979) Experience with using the Box-Cox transformation when forecasting economic time series. *J. Econometrics*, **10**, 57-69. [2.3]

Nerlove, M., Grether, D. M. and Carvalho, J. L. (1979) *Analysis of Economic Time Series*. New York: Academic Press. [4.4]

Newbold, P. (1988) Predictors projecting linear trend plus seasonal dummies. *The Statistician*, **37**, 111-127. [2.3]

Newbold, P. (1997) Business forecasting methods. In *Encyclopedia of Statistical Sciences* (ed. S. Kotz), Update Vol. 1, pp. 64-74. New York: Wiley. [6.5]

Newbold, P. and Granger, C. W. J. (1974) Experience with forecasting univariate time-series and the combination of forecasts (with discussion). *J. Roy. Statist. Soc. A*, **137**, 131-165. [6.4, 7.7]

Newbold, P., Agiakloglou, C. and Miller, J. (1993) Long-term inference based on short-term forecasting models. In *Time Series Analysis* (ed. T. Subba Rao), pp. 9-25. London: Chapman and Hall. [3.1]

Newbold, P., Agiakloglou, C. and Miller, J. (1994) Adventures with ARIMA software. *Int. J. Forecasting*, **10**, 573-581. [4.2, 6.5]

Nicholls, D. F. and Pagan, A. R. (1985) Varying coefficient regression. In *Handbook of Statistics*, vol. 5 (eds. E. J. Hannan, P. R. Krishnaiah and M. M. Rao), pp. 413-449. Amsterdam: North-Holland. [3.4]

Niu, X. (1996) Nonlinear additive models for environmental time series, with applications to ground-level ozone data analysis. *J. Amer. Statist. Assoc.*, **91**, 1310-1321. [5.6]

Novales, A. and Flores de Fruto, R. (1997) Forecasting with periodic models: A comparison with time-invariant coefficient models. *Int. J. Forecasting*, **13**, 393-405. [6.4]

O'Connor, M. and Lawrence, M. (1989) An examination of the accuracy of judgemental confidence intervals in time series forecasting. *J. Forecasting*, **8**, 141-155. [7.5]

O'Connor, M. and Lawrence, M. (1992) Time series characteristics and the widths of judgemental confidence intervals. *Int. J. Forecasting*, **7**, 413-420. [7.5]

Omran, M. F. and McKenzie, E. (1999) Testing for covariance stationarity in the U.K. all-equity returns. *Appl. Statist.*, **48**, 361-369. [3.4]

Ord, J. K., Koehler, A. B. and Snyder, R. D. (1997) Estimation and prediction for a class of dynamic nonlinear statistical models. *J. Amer. Statist. Assoc.*, **92**, 1621-1629. [3.2, 4.3, 7.5]

Pankratz, A. (1991) *Forecasting with Dynamic Regression Models*. New York: Wiley. [5.2]

Parzen, E. (1982) ARARMA models for time series analysis and forecasting. *J. Forecasting*, **1**, 67-82. [4.3, 6.4]

Patterson, D. M. and Ashley, R. A. (2000) *A Non-linear Time Series Workshop: A Toolkit for Detecting and Identifying Nonlinear Serial Dependence*. Boston, MA: Kluwer. [3.4]

Pena, D. and Box, G. E. P. (1987) Identifying a simplifying structure in time series. *J. Amer. Statist. Assoc.*, **82**, 836-843. [5.1]

Percival, D. B. and Walden, A. T. (1993) *Spectral Analysis for Physical Applications: Multitaper and Conventional Univariate Techniques*. Cambridge: Cambridge Univ. Press. [2.3]

Peters, S. C. and Freedman, D. A. (1985) Using the bootstrap to evaluate forecasting equations. *J. Forecasting*, **4**, 251-262. [7.5]

Pflaumer, P. (1988) Confidence intervals for population projections based on Monte Carlo methods. *Int. J. Forecasting*, **4**, 135-142. [7.5]

Phillips, P. C. B. (1979) The sampling distribution of forecasts from a first-order autoregression. *J. Econometrics*, **9**, 241-261. [7.4, 7.5]

Phillips, P. C. B. (1986) Understanding spurious regressions in econometrics. *J. Econometrics*, **33**, 311-340. [5.2]

Picard, R. R. and Cook, R. D. (1984) Cross-validation of regression models. *J. Amer. Statist. Assoc.*, **79**, 575-583. [8.4, 8.5]

Pole, A., West, M. and Harrison, J. (1994) *Applied Bayesian Forecasting and Time Series Analysis*. New York: Chapman and Hall. [1.5, 4.2, 6.4, 7.5]

Poskitt, D. S. and Tremayne, A. R. (1987) Determining a portfolio of linear time series models. *Biometrika*, **74**, 125-137. [8.5]

Pötscher, B. M. and Novak, A. J. (1998) The distribution of estimators after model selection: Large and small sample results. *J. Statist. Comput. Simul.*, **60**, 19-56. [8.4]

Priestley, M. B. (1981) *Spectral Analysis and Time Series*, Vols. 1 and 2. London: Academic Press. [1.5, 2.3, 2.6, 3.1, 3.2, 3.4, 4.1, 5.1, 5.3]

Priestley, M. B. (1988) *Non-linear and Non-stationary Time Series Analysis*. London: Academic Press. [3.4]

Raftery, A. E., Madigan, D. and Hoeting, J. (1997) Bayesian model averaging for linear regression models. *J. Amer. Statist. Assoc.*, **92**, 179-191. [8.5]

Ravishankar, N., Hochberg, Y., and Melnick, E. L. (1987) Approximate simultaneous prediction intervals for multiple forecasts. *Technometrics*, **29**, 371-376. [7.1]

Ravishankar, N., Shiao-Yen Wu, L. and Glaz, J. (1991) Multiple prediction intervals for time series: comparison of simultaneous and marginal intervals. *J. Forecasting*, **10**, 445-463. [7.1]

Ray, B. K. (1993) Long-range forecasting of IBM product revenues using a seasonal fractionally differenced ARMA model. *Int. J. Forecasting*, **9**, 255-269. [3.1, 6.4]

Reid, D. J. (1975). A review of short-term projection techniques. In *Practical Aspects of Forecasting* (ed. H. A. Gordon), pp. 8-25. London: Op. Res. Soc. [6.4]

Reinsel, G. (1980) Asymptotic properties of prediction errors for the multivariate autoregressive model using estimated parameters. *J. Roy. Statist. Soc. B*, **42**, 328-333. [7.4]

Reinsel, G. C. (1997) *Elements of Multivariate Time Series Analysis*, 2nd edn. New York: Springer-Verlag. [5.3]

Resnick, S. I. (1997) Heavy tail modeling and teletraffic data (with discussion). *Ann. Statist.*, **25**, 1805-1869. [7.7]

Ripley, B. D. (1996) *Pattern recognition and neural networks.* Cambridge: Cambridge Univ. Press. [3.4]

Rowe, G. and Wright, G. (1999) The Delphi technique as a forecasting tool: issues and analysis. *Int. J. Forecasting*, **15**, 353-375. [1.1]

Rycroft, R. S. (1999) Microcomputer software of interest to forecasters in comparative review: updated again. *Int. J. of Forecasting*, **15**, 93-120. [6.5]

Saltelli, A. and Scott, M. (1997) The role of sensitivity analysis in the corroboration of models and its link to model structural and parametric uncertainty. *Reliability Engineering and System Safety*, **57**, 1-4. [8.5]

Sauerbrei, W. (1999) The use of resampling methods to simplify regression models in medical statistics. *App. Stats.*, **48**, 313-329. [8.4]

Schervish, M. J. and Tsay, R. S. (1988) Bayesian modeling and forecasting in autoregressive models. In *Bayesian Analysis of Time Series and Dynamic Models* (ed. J.C. Spall), pp. 23-52. New York: Marcel Dekker. [8.5]

Schmidt, P. (1977) Some small sample evidence on the distribution of dynamic simulation forecasts. *Econometrica*, **45**, 997-1005. [7.5]

Schoemaker, P. J. H. (1991) When and how to use scenario planning: A heuristic approach with illustrations. *J. Forecasting*, **10**, 549-564. [8.5]

Schumacher, E. F. (1974) *Small is Beautiful.* London: Sphere Books Ltd. [1.3]

Shephard, N. (1996) Statistical aspects of ARCH and stochastic volatility. In *Time Series Models* (eds. D. R. Cox, D. V. Hinkley and O. E. Barndorff-Nielsen), pp. 1-67. London: Chapman and Hall. [3.4]

Smith, J. and Yadav, S. (1994) Forecasting costs incurred from unit differencing fractionally integrated processes. *Int. J. Forecasting*, **10**, 507-514. [6.4]

Spencer, D. E. (1993) Developing a Bayesian vector autoregression forecasting model. *Int. J. Forecasting*, **9**, 407-421. [5.3]

Stern, H. (1996) Neural networks in applied statistics (with discussion). *Technometrics*, **38**, 205-220. [3.4]

Stine, R. A. (1987) Estimating properties of autoregressive forecasts. *J. Amer. Statist. Assoc.*, **82**, 1072-1078. [7.5]

Stock, J. H. (1994) Unit roots, structural breaks and trends. In *Handbook of Econometrics, Vol IV* (eds. R. F. Engle and D. L. McFadden), pp. 2739-2841. Amsterdam: Elsevier. [3.1]

Stoica, P. and Nehorai, A. (1989) On multistep prediction error methods for time series models. *J. Forecasting*, **8**, 357-368. [8.5]

Subba Rao, T. and Gabr, M. M. (1984) *An Introduction to Bispectral Analysis and Bilinear Models*. New York: Springer Lecture Notes in Statistics, No. 24. [3.4]

Sutcliffe, A. (1994) Time-series forecasting using fractional differencing. *J. Forecasting*, **13**, 383-393. [6.4]

Swanson, N. R. and White, H. (1997) Forecasting economic time series using flexible versus fixed specification and linear versus nonlinear econometric models. *Int. J. Forecasting*, **13**, 439-461. [6.3, 7.7]

Tashman, L. J. and Kruk, J. M. (1996) The use of protocols to select exponential smoothing procedures: A reconsideration of forecasting competitions. *Int. J. Forecasting*, **12**, 235-253. [6.5]

Tay, A. S. and Wallis, K. F. (2000) Density forecasting: A survey. *J. of Forecasting*, **19**, 235-254. [7.1]

Taylor, A. M. R. (1997) On the practical problems of computing seasonal unit root tests. *Int. J. Forecasting*, **13**, 307-318. [3.1]

Taylor, J. W. (1999) Evaluating volatility and interval forecasts. *J. Forecasting*, **18**, 111-128. [7.1]

Taylor, J. W. and Bunn, D. W. (1999) Investigating improvements in the accuracy of prediction intervals for combination of forecasts: A simulation study. *Int. J. Forecasting*, **15**, 325-339. [4.3, 7.1, 7.5]

Taylor, S. J. (1994) Modeling stochastic volatility: A review and comparative study. *Math. Finance*, **4**, 183-204. [3.4]

Teräsvirta, T. (1994) Specification, estimation and evaluation of smooth transition autoregressive models. *J. Amer. Statist. Assoc.*, **89**, 208-218. [3.4]

Teräsvirta, T., Tjostheim, D. and Granger, C. W. J. (1994) Aspects of modelling nonlinear time series. In *Handbook of Econometrics, Vol. IV* (eds. R. F. Engle and D. L. McFadden), pp. 2917-2957. Amsterdam: Elsevier. [5.6]

Thombs, L. A. and Schucany, W. R. (1990) Bootstrap prediction intervals for autoregression. *J. Amer. Statist. Assoc.*, **85**, 486-492. [7.5, 7.6]

Thompson, P. A. and Miller, R. B. (1986) Sampling the future: A Bayesian approach to forecasting from univariate time series models. *J. Bus. Econ. Statist.*, **4**, 427-436. [7.1, 7.5, 7.8]

Tiao, G. C. and Tsay, R. S. (1994). Some advances in non-linear and adaptive modelling in time series. *J. Forecasting*, **13**, 109-131. [3.4]

Tiao, G. C. and Xu, D. (1993) Robustness of maximum likelihood estimates for multi-step predictions: The exponential smoothing case. *Biometrika*, **80**, 623-641. [8.5]

Timmermann, A. (1998) The dangers of data-driven inference: The case of calendar effects in stock returns. Discussion paper 304. LSE Financial Markets Group. [8.2]

Tong, H. (1990) *Non-linear Time Series*. Oxford: Oxford Univ. Press. [3.4, 4.2, 5.6, 7.5]

Tong, H. (1995) A personal overview of non-linear time series analysis from a chaos perspective. *Scand. J. Stats.*, **22**, 399-446. [3.4]

Vandaele, W. (1983) *Applied Time Series and Box-Jenkins Models*. New York: Academic Press. [1.5]

Veall, M. R. (1989) Applications of computationally-intensive methods to econometrics. *Bull. I.S.I.*, 47th Session, 75-88. [7.5]

Wallis, K. F. (1974) Discussion contribution to Newbold and Granger (1974). *J. Roy. Statist. Soc. A*, **137**, 155-6. [7.7]

Wallis, K. F. (ed.) (1986) *Models of the U.K. Economy*. Oxford: Oxford Univ. Press. [8.4]

Wallis, K. F. (1999) Asymmetric density forecasts of inflation and the Bank of England's fan chart. *Nat. Inst. Econ. Review*, no. 167, 106-112. [7.1]

Warner, B. and Misra, M. (1996) Understanding neural networks as statistical tools. *Amer. Statist.*, **50**, 284-293. [3.4]

Watson, M. W. (1994) Vector autoregressions and cointegration. In *Handbook of Econometrics, Vol. IV* (eds. R. F. Engle and D. L. McFadden), pp. 2843-2957. Amsterdam: Elsevier. [5.3]

Webby, R. and O'Connor, M. (1996) Judgemental and statistical time series forecasting: a review of the literature. *Int. J. Forecasting*, **12**, 91-118. [1.1, 7.5]

Wei, W. W. S. (1990) *Time Series Analysis: Univariate and Multivariate Methods*. Redwood City, CA: Addison-Wesley. [1.5, 2.4, 5.1, 7.1]

Weigend, A. S. and Gershenfeld, N. A. (eds.) (1994) *Time Series Prediction*, Proc. Vol. XV, Santa Fe Institute Studies in the Sciences of Complexity. Reading, MA: Addison-Wesley. [3.4, 6.4]

Weisberg, S. (1985) *Applied Linear Regression*. New York: Wiley. [7.4, 7.5]

West, K. D. and Cho, D. (1995) The predictive ability of several models of exchange rate volatility. *J. Econometrics*, **69**, 367-391. [3.4]

West, M. and Harrison, J. (1997) *Bayesian Forecasting and Dynamic Models*, 2nd edn. New York: Springer-Verlag. [1.5, 3.2, 4.2, 4.4, 5.6, 7.5, 8.5]

White, H. (1994) Can neural networks forecast in the big league? Comparing forecasts to the pros. Key Note Address to the 14th Int. Symposium on Forecasting, Stockholm. [6.4]

Whittle, P. (1983) *Prediction and Regulation*, 2nd edn., revised. Minneapolis: Univ. of Minnesota Press. [4.1]

Williams, W. H. and Goodman, M. L. (1971) A simple method for the construction of empirical confidence limits for economic forecasts. *J. Amer. Statist. Assoc.*, **66**, 752-754. [7.5, 7.7]

Winkler, R. L. (1972) A decision-theoretic approach to interval estimation. *J. Amer. Statist. Assoc.*, **67**, 187-191. [7.5]

Winklhofer, H., Diamantopoulos, A. and Witt, S. F. (1996) Forecasting practice: A review of the empirical literature and an agenda for future research. *Int. J. Forecasting*, **12**, 193-221. [6.5]

Wright, G., Lawrence, M. J. and Collopy, F. (1996) Editorial: The role and validity of judgment in forecasting. *Int. J. Forecasting*, **12**, 1-8. [7.5]

Yamamoto, T. (1976) Asymptotic mean square prediction error for an autoregressive model with estimated coefficients. *Appl. Statist.*, **25**, 123-127. [7.4]

Yamamoto, T. (1981) Predictions of multivariate autoregressive moving average models. *Biometrika*, **68**, 485-492. [7.4]

Yao, Q. and Tong, H. (1994) Quantifying the influence of initial values on non-linear prediction. *J. Roy. Statist. Soc. B*, **56**, 701-725. [3.4]

Yar, M. and Chatfield, C. (1990) Prediction intervals for the Holt-Winters forecasting procedure. *Int. J. Forecasting*, **6**, 1-11. [7.5, 7.6]

Zhang, G., Patuwo, B. E. and Hu, M. Y. (1998) Forecasting with artificial neural networks: The state of the art. *Int. J. Forecasting*, **14**, 35-62. [6.4]

Zhang, P. (1992) Inference after variable selection in linear regression models. *Biometrika*, **79**, 741-746. [8.2]

Index

additive seasonal effect, 21
additive outlier, 17
ad hoc methods, 95–104, 106
AEP, 102
aggregation, 11–12
AIC criterion, 77, 223–228
AIC_C criterion, 77, 223–228
airline data, 15, 81
airline model, 43
Akaike's information criterion (AIC), 77
    bias-corrected, 77
approximate formulae, 195–196
AR model, *see* autoregressive process
ARARMA method, 101, 162
ARCH model, 63
ARFIMA model, 45
ARIMA model, 35, 42–43, 217
ARMA model, 39–42, 89–90
attractor, 70
autocorrelation, 14
autocorrelation coefficient, 25
    estimation of, 31
autocorrelation function, 16, 25
autocovariance coefficient, 25
    estimation of, 31
autocovariance function, 25
automatic forecasting method, 4, 169
autoregressive (AR) process, 29, 35–37
    estimation for, 216, 220–222
    forecasting, 230–231
autoregressive conditionally
    heteroscedastic (ARCH), 63
autoregressive moving-average
    (ARMA) process, 39–42

backcasting, 22, 137
backpropagation, 68
backward shift operator, 36
basic structural model, 51

Bayesian forecasting, 93, 146, 163, 202
Bayesian model averaging, 202, 237–238
Bayesian vector autoregression, 135, 164
BDS test, 71
BIC criterion, 77, 223–228
bilinear model, 60–61, 94–95, 166
bivariate process, 27
black box, 74
bootstrapping, 198, 231
Box-Cox transformation, 17
Box-Jenkins approach, 9, 91–92, 172, 217
Box-Jenkins forecasting, 91, 162, 191
Box-Jenkins models, 35
business cycles, 14
butterfly effect, 70

calendar effects, 22, 24
case studies, 157–160, 163
causal effect, 119
causal process, 40
change points, 239
chaos, 69–71
cleaning the data, 17
closed-loop system, 111
Cochrane-Orcutt procedure, 120, 126
co-integration, 138–142
combination of forecasts, 103, 107, 175, 235
common trends model, 141, 146
computer software, 175
context, 5, 18, 81, 97, 173
continuous time series, 11
control action, 5, 13
correlogram, 16, 30–33, 75, 91
cross-correlation function, 27, 113–116, 122
cross-covariance function, 27, 113–116